Story of
Borneo Princess
and
Hurricane Irma

by John Lincoln

DORRANCE
PUBLISHING CO
EST. 1920
PITTSBURGH, PENNSYLVANIA 15238

Dorrance Publishing Co
585 Alpha Drive
Pittsburgh, PA 15238
Visit our website at www.dorrancebookstore.com

ISBN: 978-1-6393-7052-8
eISBN: 978-1-6393-7841-8

Prologue

A few days after Hurricane Irma, Mike Gasior from Stuart, Florida, an experienced first mate on large tanker ships on his days off, and a buddy Aaron Wallace went out on a friend's 27-foot Boston Whaler—*Freaky Sabiki*, owned by Jeremy Lund to make a run offshore to see if there were any fish around.

The storm had only passed a few days before, and even though the waters were all churned up, they figured it was still worth a shot. They hit St. Lucie inlet at around 0730 and would run until they found good marks or weed lines. They were barely a quarter mile offshore and the debris fields began, and they had to dodge everything from the complete frame sides of people's houses to capsized vessels. It was literally a minefield of debris. They finally made it out to about 500 feet of water approximately 6–7 miles offshore and decided to troll around the massive weed beds.

After several hours with only a few fish in the box, they picked up and ran back inshore to the debris fields. Mike later would say it was insane to see the amount of loss just floating by, some items were even adrift about 20 feet below the surface just wasting away. They came up on a huge dock approximately 100-feet long and drifted next to it just to see if there was anything around.

They found peoples' clothes, books, and thousands of lobster/crab pot buoys out there. They claim they could have filled up the whole boat and more!

Finally, it was about 1600 and they were going to make the run back to the inlet when they stopped at one more huge patch, so big that Jeremy backed up to it to blow some of the weed away with the engines so they could make their way to the middle. After a few minutes, they made it inside this 1/2–3/4-mile-wide debris field. Aaron was hanging over one gunnel, Mike over the other, and Jeremy was at the helm. Mike then noticed something just bobbing in the middle of the debris / weed with an antenna on it. One of the guys thought maybe a small weather beacon and as soon as they got closer Mike said, "No that is an EPIRB!!" Jeremy got Mike close enough to grab it and as soon as he did, noticed it had already been activated for longer than ninety-six hours the normal battery life. Mike immediately called his wife Alison whom was also a merchant mariner on container ships and told her to call the Coast Guard.

She did just that and luckily the EPIRB had the HIN on it, which she gave to the USCG. They were able to advise her that the vessel did go down in the storm and that the owner would be notified that they had found his EPIRB.

Contents

The Borneo Princess

1.0

The Borneo Princess

1.1 Leading Up to its Development

In 2010 OES Limited, the company owned mainly by John Lincoln however 0.5% of the company or 150,000 shares was given away each year to long-term employees of ten years or more. Bill Lincoln, Sean Lincoln, Melissa Lincoln, and Harinder Pal Singh all qualified for this.

In 2010 the company had experienced some great previous years. It had picked up a 40-million-dollar contract in Australia in 2009 and had completed what we thought was the difficult work and assumed we would be paid for our work, and had also completed early in the year a 3-million-dollar pipeline trenching project in Angola for which we had been thoroughly paid. I had money and decided to spend it on new equipment and new ventures.

Chris Foreman, who worked for the company, and brother Bill Lincoln both had Indonesian wives and lived in Balikpapan and OES had done much work in previous years as a foreign specialist company, and so opening an Indonesian Company seemed a good thing, especially when Chris found a new five-acre shipyard with a 2000-square-foot shop for rent from a company for only USD 50,000 per year. The same facility in Singapore would be USD 500,000 per year and besides we were a lot closer to projects in Balikpapan, the Philippines, and Papua New Guinea.

Little did we know the awaiting perils and outrageous taxation, dishonest Indonesian companies, and government bureaucratic red tape we would find trying to do almost anything. For example, later we would find out we had to give half the company away to Bill's wife, Sari, in Indonesia, and to import a barge, we would buy in Singapore there was a customs duty of 50% or USD 500,000, which we never would pay.

It was time for OES, still naive to the ways of Indonesia, to expand and be a pipe-lay contractor and trench our own pipelines. At least we would pay ourselves, which was always a problem I thought was confined to only third-world countries. We signed a five-year lease with an option for five more years. We bought a 230 ft. x 60 ft. heavy deck cargo barge, which we planned to convert to a pipe-lay barge. I also designed an 80 ft x 33 ft. wide steel catamaran motor sailor in concept only and got a very good price from a reliable steel fabricator named Davidi in Balikpapan that had successfully built our Buya Besar (Big Crocodile) steel catamaran construction barge for us before. The price included steel welding, blasting and painting, and interior work. We were to provide all mechanical, electrical, instrumentation, and sailing facilities.

1.2 Design and Concept

OES—having a very broad experience and also very innovational in the offshore Oil and Gas business on many different types of vessels, a general love for the ocean, combined with the personal interests of John, Sean, and Bill Lincoln involved in pleasure diving and sailing—resulted in the Lincolns developing a new hybrid of a vessel, a motor sailor catamaran yacht, which serves many roles, including a shallow water dive support, survey, crew boat, transport ferry, fast naval small craft, and international sailing vessel.

The vessel concept was also, sticking to the modus operandi of John Lincoln—a nonstop innovator—would be a world first of its type.

It was decided to limit the length to 24 meters and can therefore be captained by the owner or owner's representative in most countries without any special captain's certification. A Catamaran twin hull was selected on the basis of stability, space, and comfort. A bonus is the hull speed, the maximum reasonable speed of a vessel before a huge power must accompany an increase in speed associated with the vertical climb angle riding up the bow wave of a catamaran is much more than that for a mono hull as one hull rides the bow wave depression of the other. The hull speed is 15 knots. A width of 10 meters was selected, which provided two 2.5-meter-wide hulls and state room dimensions spaced 5 meters the width of the main entertainment and dining area.

A steel hull was selected on the basis of strength, reliability, and the ease at which attachments can be added or deleted by welding. Aluminum is not a reliable material as it ages; it weakens, cracks, and undergoes excessive galvanic corrosion in very salty water like in the Middle East and in the presence of any steel. In addition, aluminum vessels bounce around too much and are noisy and conduct too much heat away. Most Oil Company and Marine Classification Societies also require steel hulls.

With steel the only drawback is ugly red rust, and so a very comprehensive sandblasting and marine coating system was adopted. A thick coating of inorganic zinc silicate, which acts like an anode and also galvanizing was applied followed by epoxy topcoat and polyurethane final coating. Many other components like the mast, boom, and stay wires are all hot dipped galvanized. No aluminum at all is used except for the window frames and they are heavily insulated from the steel.

The *Borneo Princess* was designed to qualify for ABS classification for up to 50 miles offshore for commercial operations. Windows are all thick shatter-proof glass and slightly tinted for heat conservation. There are three water-tight doors and two emergency escape hatches for the front bedrooms. In total there are ten water-tight compartments such that if a main one is flooded on one side or two for both sides the vessel was not supposed to sink. This was in theory only as there were some manufacture errors that must have allowed water to pass through the bulkheads.

It has a double bow as a result John did not like the original bow; so it got extra safety. The below-water hull is 3/8 inch thick for safety and long life. The 16-inch-deep continuous keel is sacrificial which means it could be broken off say if a reef was hit and no damage to the hull would result. The propellers are not only in Kort nozzles and idea John got while on a Cruise liner voyage to Tasmania in 2009 but are recessed under the hull again safe from any collision with a reef. Each pontoon is fitted with five water-tight compartments, so if there was a collision that opened the hull below the waterline it would not sink. Two unplanned collisions in the ship yard proved the durability and flexibility of the hull steel and welding. It has a self-draining teak stern deck, IPO commercial shipping rated unbreakable windows and a single-entry water tight lock door into the wheel house.

Add two life rafts an EPIRB emergency beacon, fire extinguishers and a sea anchor it is pretty safe in rough seas. Unlike a sailing vessel twin large reliable, powerful and fuel efficient 450 Horsepower diesel Caterpillar engines were added for reasonable speed and power. The cruising speed is about 10 knots. The *Borneo Princess* has four fuel tanks with 15-tonne total capacity allowing a range of about 4000 nautical miles. Fuel can be easily transferred by pumps to either side to assist with ballast while under sail. The vessel has twin 17kw Caterpillar diesel generators a robust central air conditioning system with water cooled condenser and 700 gallons per day fresh water maker. Ten tonnes of water storage allows extra ballast that can be transferred from side to side for fast sailing.

The vessel as a sailing vessel is a sloop and is designed to sail in 40-knot winds with both sails fully out and without opposite side ballast.

The 24-meter-high mast and 12-meter-long boom are high tensile steel and hot dip galvanized. The wind load under this condition is 20 tonnes and the vessel list approximately 9 degrees. A safe limit of 7 degrees was imposed however.

It had stealth capability as a means to reduce wind loads on the sides or from the bow. The mast and boom and sailing systems were structurally designed for 40 knot winds side on with a factor of safety of two. The provision of ten tonnes of fresh ballast water in Stainless Steel tanks could be transferred in under an hour to port or starboard or fore and aft for stability or trim. The design of the bow unlike most catamarans makes the bow come up from water that passes under the hull.

The hand rails, bollards, winches, sail rigging, ladders and fuel and water fill points and vents were Stainless Steel giving it more of a yachting look. The right-angle corners were all rounded off again giving it a rounded yachting look. The corridors around the boat were wide enough for a wheel chair to navigate around. The welds on the sides and saloon roof and deck were full penetration ground smooth welds again giving it a yachting look. The back deck, transom and interior floor work were teak giving it that special elegant look combined with the strength and safety of commercial ships.

There are two large state rooms with Queen Beds and four two-man bunk rooms. There are two large bath rooms with showers and toilet and hot water. A galley is equipped with a four-burner stove, double sink, large refrigerator, freezer, microwave, dining table and walk in storage pantry. The main living and entertainment area features a twelve-man dining table, a bar, four couches and lounges, reading lights and a 48-inch TV screen with hundreds of DVDs and a reading library.

The vessel comes with a comprehensive "state of the art" instrumentation and control technology. It is outfitted with DGPS, satellite phone and internet, weather reports, SSB and VHF radio, sonar, radar and an automatic pilot system with alarms for possible concerns with other shipping traffic or changes in water depth. For additional safety a sixteen-man life raft and thirteen-man inflatable boats are also supplied.

A sextant, chronometer and paper charts provided in the event of failure of DGPS.

The basic dimensions, hull thickness and weight can be changed for any future vessels without any fundamental departure from the design concept. For example, 2 x 1200 HP engines and a lighter vessel, say 25 tonnes instead of 50 tonnes, will allow a speed of almost twice or about 30 knots which may be desirable for crew, military or ferry boats. The width and length can be increased up to 20% for those that can afford huge comfort and desire a modified floor plan.

What we did not know but would find out the steering system for the Kort nozzles was way undersized and the port side shaft propeller connection to the engine coupling was not right.

2.0

The Early Days in Balikpapan 2010 to 2015

In 2010 we leased for five years a four-acre property on the water and a 2000 square meter covered work shop / warehouse from a local company named Intrapatama run by Muslims. The concrete wharf was not yet completed but there was a 40-feet-wide concrete ramp.

We let Davidi, a company run by Christians that we knew well over the years use our work shop for fabrication projects. They were very good welders and Indonesian steel very reliable in properties of strength and ductility. They had built catamaran work barges for us dating back to 2002 when we had a big oil and gas pipeline burial project in the area. They also had good detail designers. We gave them an EPC contract to design and build the *Borneo Princess* steel work and interior based on only pencil drawings by myself of layout, elevations, cabin floor plan, structures an estimated steel weight of 75 tonnes and general welding, material and painting specifications. They built the vessel in the front of our yard. We used our 100-tonne crane we bought and shipped from Jakarta and it drove off the ship and up our ramp.

Through myself, American Chris Foreman as site manager, Sari as part time director and her two brothers Dion and Dodi, another employee of ours Ryan as mechanical workers we supervised Davidi and undertook all mechanical, electrical and instrument procurement and installation. Sean Lincoln and Bill Lincoln busy on other projects would get involved from time to time. Sean would be responsible for all sailing and rigging systems a huge task. We also hired Astrid a young Christian college accounting graduate that would be invaluable as a secretary making sure we followed all of the required Indonesian Government income tax reporting, payroll and labor law requirements.

We rented a very large five-bedroom two-story house on the beach at first which was very economical about USD 12,000 per year. Other OES employees like Sean, Vanderley and Patrick would visit and do some work when we had no normal offshore trenching for them. Patrick, I met in 2012 he was my plumber in Key Largo and Vanderley was the brother of my second wife Maria whom lived in Australia and we were divorced. Vanderley had worked for us since 2000.

The area had two places of interest for us Jacks Bar virtually next door on the water owned by an ex-Louisiana welder named Jack where you could get American food. The other was Lilas, a bar also on the water with bar girls to cater for your every need whom had cheap beer and a friendly atmosphere. Most expat Australians and one Swede would frequent the place after work every day. Chris was an alcoholic addicted to beer and a chain smoker which gave him little patience I lectured him constantly as he drove like crazy through Indonesian motor scooter traffic swerving and yelling at the riders. I really worried he would kill someone, and so he was not allowed to drive a company car. Chris finally quit in late 2011 and this was not a problem as Bill Lincoln my brother had joined us and easily took over Chris's work. We did have a good time with Chris and his wife Mary when we visited.

Batit and I not yet married would travel together and visit Balikpapan for a week or two about three times a year. The house had a reputation of being haunted as there were still many Japanese bunkers in the area used during World War II and many Japanese died. Vanderley apparently had seen a woman floating around the room. Yusi, Sean's girl friend stayed for a while and also had weird experiences. The place was so big Batit and I used to joke around that we had to take a taxi to the kitchen. One night, while just going to bed we heard a series of very loud bangs coming from the ceiling and we jumped up and assumed somebody was breaking in from the roof as the windows all had bars. I ran out on the porch and started screaming for help.

Batit got out of bed and the floor was up higher where she stepped and thought we had an earthquake and ran downstairs and outside. I discovered what she was saying and turned on the bedroom light to find a large ridge of cement tiles that had popped up a half a foot all across the room. I knew then we had no intruders but what had caused this. In the meantime, Batit had run out side out of the back door bare foot and stepped on a snake we believed to be a cobra snake. She ran back inside with the cobra following her under her feet probably as scared as she was. She was not bitten. We had to go to a hotel that night still puzzled if there was some supernatural power at work here. Finally, I settled on foundation settlement as the cause convinced her and we went back the next night.

Vanderley had quite a reputation as a ladies' man, reminding me of myself in earlier years, picking up two or three a week and taking them back to his room in the house. One girl whom he gave a key to came there one night and trashed the place.

In late 2012 Bill was on a project in Singapore, and so Patrick and I were in Balikpapan together and Batit stayed at my apartment in Melbourne. Patrick and I flew to the island of Manado, a fabulous dive site, one weekend and did wall diving around a volcano and you could hear the volcano rumble underwater which was very eerie. Upon our return, Patrick was busted by the Indonesian labor board as he did not have proper work papers. They left me alone. The fine only $50 as Astrid came to our defense as good employers.

In 2013 we had a huge falling out with Intrapatama as they wanted extra money because we had subleased an acre out the property to a company wanting to store some equipment. They were also in the process of completing the concrete wharf our wharf and when we took our crane there, they went crazy saying the wharf would not support our crane and at one point sent guys with clubs around in case we tried. They claimed our contract was not valid theirs in Indonesian language was. This was my first exposure to lying cheating Indonesian bastards. We decided to move to another ship yard called Meranti ship yard and there we finished the boat. Bill and Meranti and a Singaporean guy named David Soh would also walk the boat across the yard on large inflatable rollers, down the ramp and launch it in the water the first time. It floated much to Bills surprise obviously not having much faith in his brothers engineering ability.

Meranti were Muslims but honest and good to deal with. We had to complete the sailing facilities, electrical work, painting, interior and other things like A frame, engine room air intakes and vents.

The work was going very slow as most of the facilities required much design work. It is said the last 10% of every project takes 80% of the time and we were finding this to be true.

All subcontractors had run out of money the work was more than they had allowed for. We paid extra money not much maybe 20% more to electrical, hydraulic and general contractor Davidi. Things would also slow down as Bill and some of our guys would disappear in parts of 2012, 2013, 2014 and 2015 working in Singapore, Israel, Ghana and Mexico.

3.0

Completion of Vessel

In December 2015 I hired two experienced captains American Tomas Simonson, a nuclear submarine weapons officer and whom had also just sailed his personal sailing yacht to New Zealand from Seattle and an Australian named Kevin a very experienced all around commercial and sailing captain. Tomas and Kevin conducted a safety and sail ability audit of the vessel and a ninety-item check list was established.

The next seven months were spent making the necessary changes as well as installing an auto pilot system and commissioning all the instrumentation systems largely undertaken by Tomas and assisted by Sean Lincoln.

In March 2016 while in Key Largo sending brother Bill a new Air Conditioning thermostat from the UPS store I bumped into Joe Dyll an experienced motor and sailing captain whom also worked part time as the Dock Master for the Key Largo Marina del Mar. Joe helped explain to me and helped with making final decisions of the recommendations suggested by Tomas and Kevin like for example the auto pilot system. Joe helped in the voyage studies working out speeds based on probable wind data and vessel under power to determine the necessary stops and fuel costs. In the beginning a trip around Africa was decided as a safer and quicker route rather than through the Red Sea a heavily pirated area, east in the roaring forties below Australia probably too rough in following seas, and east across the Pacific with little winds and far and few fuel availability locations.

About half way through the study, it was discovered by Joe it was much less expensive and far less risky to just go to Singapore where we could be loaded on a Super Freighter and take us safely to Fort Lauderdale near home sweet home.

4.0

Sea Trials and the 1100 Mile Maiden Voyage to Singapore

4.1 The Journey of John and Joe to Balikpapan, Borneo, Indonesia
John wanting to close the Indonesia office and yard and seriously cut expenses and after thousands of emails and decisions over a sixth month period at the advice from Tomas everything should be ready to sail to Singapore by August, John and Joe board the first of four planes bound for Singapore with cheap tickets John got on the internet. Joe and John had already arranged passage of the *Borneo Princess* on a United Yacht Transport container ship from Singapore to Fort Lauderdale that would also first visit Taiwan, Canada and then transit through the Panama Canal with the entire voyage taking three months.

John's wife Batit and Joe's girlfriend Marta also the drivers accompany them to the Fort Lauderdale airport. The departure is at 5:00 p.m. to Atlanta then catch another flight to Detroit spend the night in a hotel then catch a flight to Tokyo the following morning. John the seasoned traveler whom used to do over a hundred international flights a year is just waiting for something to screw up the plans which seems to always happen on his flights in the last few years so is being very careful. At the Airport at Fort Lauderdale plans to have dinner with Batit and

Marta is quickly halted as the airport has no restaurant until after security. Joe wants to use the black porter express check in outside the airport. John is hesitant at first but we were only checking bags to Detroit. The porter sees the bags are slightly overweight but says nothing so I tip him $20 at the request of Joe.

The plane to Detroit is boarded after a few hours then shit starts to raise its head. The check in clerk screwed up there was a wrong head count probably one of his friends was let on for free and everyone had to get off the plane and re board with care this time. Just as well some of the people it was discovered were actually on the wrong plane. Well, when we get to Detroit at 1:00 a.m. not only is it an hour later, but the baggage handler is broken and we can't get our bags. Finally, I go down the luggage chute others follow and we start to retrieve our bags by hand and slide them down to people.

After that trauma, it's now 2:00 a.m. I ring the hotel not knowing I was given the wrong phone number on the flight itinerary of the correct hotel chain but in another city and ask them to send the courtesy bus. They say they sent it but no bus I keep ringing until I finally learn of their screw up. When I get the correct number for Detroit the desk clerk tells me the shuttle bus stops after 1:00 a.m. We go out and get a Muslin taxi driver for some reason whom I just did not trust when he said $50 to the hotel. I quiz him about ISIS in the Middle East, and he says he never heard of them and I ask him what he thinks about young men conducting Jihad killing hundreds of innocent Westerners.

He said it was not true was made up lies. I am ready to punch this asshole but decide on a more peaceful existence and pay his outrageous fare. We get to the hotel, some thirtysomething young men has got this hot-to-trot young prostitute in short ripped jean shorts and high heels and is impatiently trying to check in, but we just beat him to the counter. Two prostitutes come down with a large black guy and I am thinking a real shithole.

The desk clerk was totally confused could not find our booking and blamed cheap ticket dotcom for the screw up with the wrong phone number. He then wants payment and I said we already paid for the rooms. Then I pay again and he gives us one room to share. I yell, "We want two fucking rooms; it's 3:00 a.m. We are not here for gay sex!" He tells us the shuttle runs at 8:00 a.m. the following morning, so we book it.

Well up to airport get great flight sit in back row of plane seats actually go back unlike most planes. There were two empty seats between us so we had room and comfort. I never fly Delta the food was crap and for an international flight the entertainment movies and music was limited crap as well. Then after a quick meal the next meal is a small slice of pizza six hours later then nothing for six more hours.

We get to Tokyo the flight to Singapore is delayed so we get on one an hour earlier. Finally, we arrive in Singapore my favorite city early Sunday night and within thirty minutes we are away by taxi to our luxurious hotel rooms. After twelve hours of uninterrupted stretched out sleep I stumble down for breakfast and bump into Joe. I arranged to spend two nights in Singapore to recover and also do some banking that is get $10,000 cash for the trip, go see Aqua Marine and bitch about the crap inflatable boat they sold us and advised we would return it when boat arrives for the correct white one they promised in their email, but of course that was four years ago. We did Sim Lim Square for lunch and walked to Bugis junction where I got a Singapore sim card for my I phone and then walked back to the hotel down Queen Street.

That night we hit Orchard Towers or also called "Four Floors of Whores" and were hit on all night while having two beers by young Asian girls wanting to go home with us for a few hundred dollars. We did not partake too tired, maybe too old in any event not interested. Their prices dropped to less than a hundred dollars still we said no. If I was a horny Jew, could have scored for $75 I am sure. I missed my wife there was no substitute for her besides disease and money concerns just not interested.

Next day boarded a flight to Balikpapan via Jakarta because we wanted to smuggle the sea anchor in as Balikpapan never checks domestic flights and Jakarta far too busy to worry about two gringos with a cardboard box as luggage. A great route for deception has smuggled $80,000 worth of instruments in this way but the ordeal is an all-day affair and two flights but cheap $100 each way. We arrive are picked up by Sean and taken to the $500 per month OES 3- bedroom, 4-bath guest house / slash office. Tomas had been shacking up with his nineteen-year-old girlfriend Sella but vacated room with king bed for me and I did not know it for a while but they were sleeping together on a single bed in the other upstairs room.

The house is in Balikpapan Baru, a neat little village with beautiful two-story concrete homes. The shopping areas in the neighborhood look like something out of Disney World with towers and unique architecture everywhere. Problem is it is an energy town with income from the many coal mines, and oil and gas fields all having financial problems as the prices of energy commodities has collapsed. Many homes have been evacuated as the expats that were here have left.

The biggest Oil Company Total has sold its business to Pertamina the state-run Oil Company. Our house is sort of neat all white tiled floor with a large tiled staircase with black wrought iron banister and teak hand rails. There is a bed room downstairs and a western style bath just down the hall. In the back there is a small room the maid quarters with a standard Indonesian toilet a pan in the floor. Upstairs there is a large bedroom with a bathroom built in and another with a bathroom in the hall. All bedrooms have air conditioners. Only one bathroom has hot water and is upstairs so showers are usually with cold water. Water is pumped from a tank or from the city. When the power fails about three times per week the pump fails and only tank water will gravity drain for downstairs only, provided you remembered to fill the tank.

There is a diesel generator during power failure you have to start provided there is fuel and often it runs out. With no power I have no phone, computer, internet, TV, air conditioning in the 90-degree heat or water sometimes and usually there is no car here as Sean and Thomas have taken to the boat or else Astrid is running errands. Once I walked next door to Sari's house, Bill's wife, a director who was too busy to take me to get fuel, let me use her hand phone.

5.0

Borneo Princess Final Preparations and Government Approval

Next day we were off to see the boat. I had not seen it for 1.5 years just followed and directed the progress on email. They were still painting final coat on top deck and had almost finished the teak back deck. Some large boxes to be used as seats and storage we finally rejected on the basis that they were made of veneer and it would peel away and were too high to sit on with legs dangling. I mean I was sorry to see them go. It took two years to get them and during the same time to get the teak deck after two unsuccessful attempts one in which they poured concrete on the stern deck, and I screamed, "get that fucking concrete off my already too heavy boat" and then proceeded to design the deck and attachment method for them.

Still no teak transom. It was a great looking boat though I had never seen the interior wood work or the mast and boom and sailing winches and facilities. I tried out the toilets, room lights, refrigerator, and windows anything to get familiar. I was impressed with everything including large stainless bars to help up and down the three small staircases into and out of saloon. I also noticed the stainless A frame to support the tarpaulin over the back deck.

Next day Thursday we put up the jib sail and installed the furler and Jib sail ropes. Everything fit like a glove a great tribute to my son Sean that designed it with me actually doing the concept for the furler involving a swiveling 3-inch HDPE pipe over the jib stay 25 mm wire. The jib sail was fitted with a zipper and was zipped over the 3-inch HDPE. Everyone seems to accept and like it and a fraction of the cost of an aluminum one and you don't have to worry about galvanic corrosion between the steel wire and aluminum furler sheath.

Tomas still busy with electrics and showing Joe how boat operates and not everything working. I am more concerned with possible commercial stumbling blocks say if Davidi our boat builder or shipyard is owed money and the permit with the government to leave Balikpapan. Getting boat out was "a real can of worms" the government never heard of a private yacht being built in Indonesia much less one with a foreign flag and wanting to leave so no one knew what to do.

How do we screw these people and get their money is what I thought they were thinking. An agent Edo was irritated with Tomas for contacting the Port Authority direct and at one point told him to fuck off but I was able to get him inline by getting his client and my friend of ten years Billy Towul to arrange a meeting to discuss options. Every agent and Indonesia government Customs, Immigration and Port Authorities were dumbfounded and seemed irritated at their inability to handle the departure of the boat. Edo advised his fees for the cruising permit was 40 million rupia or USD 3,000 which I thought was high. At one point I told our people we were going to get an easy sea trial test permission that was good for three days and in two days we would be out of the country in Malaysia if we headed north and never reenter Indonesia on our way to Singapore. Singapore does not like Indonesia and purposely do things to piss them off. Friday much of the same of the previous day then knock of early as Tomas is getting married Sunday to Sella. Saturday, we drive in two cars Astrid our Christian office girl in hers and Joe and I with Sean in company car or small SUV four hours to Samarinda.

We check into a small hotel on the river good clean rooms for USD 35 per night and Sean and I swam some laps in the pool. Later at night we all went into the city to a high security club restaurant on the penthouse open roof top floor and if the food was not great and the beer too expensive it was at least cool as an ocean breeze blew in.

Next day Astrid came with us and on the way to the wedding at the brides house an hour north and we all got lost in the mountains for hours as neither her or Sean's GPS seemed to work. Luckily Joes GPS from the USA worked fine and he steered the course correctly. We got there and we were the only people except two of Tomas friends from Singapore and Sella's older sister Linda and her Austrian husband Robert that spoke English. Linda was a real dry humor comedian asking Sean and Joe if they were lovers since they headed for the toilet at the same time.

We did not drink knowing we would have a long dangerous five-hour drive back to Balikpapan but had a really great time talking and eating really great food.

Next day, Monday, Joe and I talked to Wimba the Meranti shipyard owner's son / manager because at the time we all believed the boat had to be treated like a commercial vessel and certified and the builder's drawings had to be submitted Plan A. Davidi had fallen out of favor with the Port Authority as Davidi had a boat that recently wrecked, killing three other people, so he was going to use Meranti's name. Wimba wanted USD 2,000 to use his company's name and logo on our drawings Plan B.

Sean and I flew to Jakarta for a meeting with Technip on Tuesday and had to stay the night as the airline Lion Air screwed up the flight schedule on the board showing a later flight then flew early causing us to miss our flight arriving late Wednesday and two days lost for us on the boat.

We had a meeting finally on boat with Billy of Davidi, our agent Edo and Astrid. They did not want Tomas or Joe in the meeting in the beginning wanting to first smooth over the political approach. Redo was now open to pursue the private yacht approach first discovered by Tomas. Treating it like a private yacht that had entered Indonesian waters as boat had potential problems as it was built in Indonesia. Joe and Tomas walked into the meeting late and did not know Edo had agreed to pursue the new private yacht regulations and also agreed to get us a sea trial permit asap Plan C.

In the interim I wanted to test this boat for six years it had never been operated so pushed for that. A sea trial permit could easily be granted according to Edo. The sea trial permit was promised by Thursday night as I wanted to do sea trials urgently but did not arrive. During a Friday lunch an argument broke out between me and Joe as he expressed pessimism at getting the permit and I told him we would get it and that there was to be no more fucking pessimism. We got it that night and next day Saturday we were off.

Joe as captain was sort of left alone to do his thing in the 3-knot river learning how the boat turned and having problems wanting little advice from me. Problem was he was trying to steer the boat with the engines but at the same time turning the Kort nozzles to one side which only causes them to cancel each other's effect. He finally found his own method and maneuvered until he was happy and headed out for open water. The boat seemed to hold up well in 6 ft seas except one wood platform that was clamped came loose. Tomas had put 4 tonnes of water in the bow that we did not need so I had it removed but we quit testing so did not get the results from this. These panels look like they could easily be fixed by strapping each other together with rope and to the two steel beams.

The biggest problem seemed to be excessive vibration above 1350 rpm so we limited it to 1350 rpm. At that rpm we achieved a speed of 6 knots. The props would be if necessary reengineered in Florida and vibration problems sorted. Selection of the props was by me and I probably had made a mistake. I would find out later we needed more displacement or pitch and would get more torque. A flat end prop was suggested by Joe. The extremely good news was that the fuel consumption was very low only five gallons per hour per engine. This means in the USA I could carry a little fuel enough for forty-eight-hour cruising a little water as I had a water maker and boat would only draw 5 feet of water allowing me to park it behind my house even at low tide.

Later Joe navigated well coming down river against the current steering between the parked barges and tugs. When we returned to port, we found a foot or so of water in the starboard bilge and the bilge pumps did not work. Water was discovered to come from the wet exhaust system a three quarter inch nozzle connection with a valve appeared to rupture at the branch. The fittings were found to be of weak TP 304 material and it was also a weak connection with no reinforcement made worse by the long valve handle. The port side had already been reinforced previously but they forgot the starboard side. We welded in a new coupling and the valve replaced by a threaded plug.

Sunday and Monday were days spent on the boat with Joe and Sean installing the mainsail battens. Tomas working on the AC system, navigational lights and bilge pumps. I went shopping for supplies, tools and non-perishable food. We were only on shore power which we could not rely on as people would accidently disconnect it or else a city power failure. We could only run generators at high tide otherwise suck mud into the cooling water intakes. We had 5 tonnes or 1330 gallons of fuel delivered costing us about USD 3,500 or about USD 2.60 per gallon.

On Wednesday I boarded a flight for Singapore to have a meeting on Thursday with HSBC whom was threatening to close or account as they thought we had unusual banking habits and were suspected of smuggling drugs. Idiots I sorted them out one idiot nosey black eyed 30 something skinny bitch named Jean was behind it all as she did not like our email response a year previous when I told her questions were silly. I also went to bank transferred money into business account from my personal and picked up two larger bilge pumps at Aqua. I returned Friday on the direct flight to Balikpapan on Silk Air. Sean had been detained by immigration in a random check at his hotel because he was wearing coveralls. I was amazed to find Joe had left to make his way back to the USA without thoroughly advising me of the timing.

Tomas and Sean also went to Singapore on Saturday as the latest constantly evolving requirements to go to Batam required them to re-enter on business visas Plan D. I had a relaxing time by myself in the house watching movies and boxing matches, shopping for food, and doing laundry. And, of course, I spent hours each day with Batit on video Skype something I could not do in Singapore hotel just too choked with Wi-Fi traffic. On Monday business as usual we went down to boat. Tomas and Sean were busy getting sailing rigging ready and testing the AC system and trying to sort out the bilge pump problems and some remaining navigational light problems. I get a call from the agent Customs is concerned and I start to worry these bastards want to tax us on exports same thing which started the American Civil War. So, I worry but trying to be positive and have a backup plan get the fuck out of Indonesia with boat fast. I would get Astrid to book a meeting with Customs and the agent and her and I.

On Tuesday we prepare a commercial invoice which shows PT OES (Indonesia) selling the boat to OES Limited for one dollar a normal thing in the Western world if both companies are owned by same person Plan E. It did not fly well but did get them thinking. Luckily, we had a mentor whom worked with Edo in the form of sexy black veiled Muslim women named Melie whom wanted to help us and was very influential dealing with the male bosses whom were not familiar with our unique case. Astrid also told them how we opened an office for six years never got a single job in Indonesia but supported the lives of five Indonesians and Subcontractors there building the *Princess*. That did it they in theory approved our departure of the first private yacht built in Indonesia and also foreign flagged. I guess my Karma was good as I had been defending Muslims for years against many westerners whom don't understand most 99.99 % are good.

Tomas gets back bad news AC is fucked burnt out somewhere between the time when we tried to test it and it got hot from insufficient cooling water the day before to previously when the shore power shorted out and fried some wires. No problem John and Astrid had already bought fans as a backup. We then get a call the Batam Port Authority was not so convinced so had to do a commercial invoice from Davidi whom did most of the construction Plan F. Davidi had no problem with that I promised to send them USD 30,000 as a bonus once we got out of Indonesia and they trusted me and I will do it no problem they deserve ten times that for all the variations they never claimed. Once we know how to get out, and now we know how to build a boat and exit again.

On Wednesday 31 August, Sean, Tomas, Dion, and Dodi and five David people finished installing the main sail. Astrid and I continue shopping for supply's and get some deep-water trolling fishing tackle. Thursday was to be the final day of preparation tying down the wood grate tarpaulins, getting perishable foods and notifying every one of our satellite phone number they could text for free. Friday 2 Sept we would be off at 6:00 a.m. we hoped but shit would happen again.

We got a call from the agent Edo evidently Davidi export license had expired and we needed to work with another company, and so another meeting was arranged for 2:00 p.m. at customs and to meet this new company. I knew it was too good to be true bureaucratic constipation again. A new angle to screw us now Plan G.

Edo arranged a meeting with a friend of his, a woman named Keekee who ran a shipping line and had an export license. After several minutes in the meeting, she agreed to help us for 25 million or USD 2,000. I tried to get her down to 10 million then 15 million, but she would not budge. It was also agreed Plan H to change all the documents to show us going to Singapore not wanting to repeat all this bullshit a second time. All passports would be stamped out of Indonesia except Sean whom would get off in Indonesia Tanjung Pandan catch a flight to Samarinda to join his wife Yusi visiting her mother. Then we get a call the Port Authority in Balikpapan is not allowed to clear any vessel to Singapore. So, it's back to previous plans G clear to Batam then clear to Singapore a week or so later so we thought.

The next day Friday we learn Keekee's boss will not allow us to use their export license but Edo and Melie found another company license we could use. This changed all the documents and Melie went down and did her woman thing of smiling and shaking her nice boobs and ass in front of the Customs officials. Astrid and I spent the day shopping for paint, two stroke oil for the outboard dingy engine, transmission oil for the main engine gear boxes and other provisions. It was stressful a lot of traffic could not easily find either oil. Then she springs it on me that it was September and we needed to pay for her mother's car rental for another a month. I explain we will be leaving soon, and so then she wants a daily rate almost twice of what we had been paying, I really lost my temper and explained we had been renting her mother's car for six years and her mother could have bought two new cars for what we paid in rent why try and screw us now. She got upset and had tears.

She also made the statement what was wrong with this company why Tomas did not order the gear oil before. I said "Astrid you are the company you and Bill were here for six years why didn't you order the oil." Thomas was sent to only do the instrumentation "The transmission oil they would only sell us a 200-liter drum but we could not handle it." Finally, we rang Gooday the Caterpillar dealer they had some in smaller 20-liter containers they wanted twice the volumetric price. It was late we had no time to drive to their place at another city so asked Tomas to go there. They would not let us have it on credit how to pay. We paid USD 250,000 for engines and generators yet they could not trust us for USD 200 worth of oil.

Finally, Astrid showed a bit of Western ingenuity she could transfer from her personal account via online banking through her smart phone. Late Friday customs rang Melie they would need something else a letter signed by boat owner. as on official registration documents for OES Limited I was 100% owner of PT OES and OES Limited, that this boat was not for commercial profit but for my personal use we could submit it in the morning.

That night we returned to the house to find a large barbeque being thrown by the next-door neighbor an alcoholic fifty-year-old Texas Oil man named Stacey whom worked for a local Indonesian oil company. He shouted over the fence "Come on over plenty of beer and food" so we went. Great ribs and chicken, baked style red beans and garlic bread just what I needed I had skipped lunch and was tired of eating road kill chicken.

I had three plate loads. Stacey lived with his Indonesian girlfriend Mary a thirtysomething sweet intelligent street wise pretty girl. The had another couple show a fifty-year-old semi-retired, semi unemployed oil well consultant Canadian man Roger and his thirty-year-old something blonde Chinese looking girlfriend. They were all great people. Stacey's house had three refrigerators with one in the living room two were for beer and whiskey.

All night I am thinking why does customs want this letter for what reason does it mean if we have a commercial interest say we want to manufacture and sell yacht's is there an export duty or just for paperwork bureaucracy. Next mornings I awake around 9.30 a.m. I Skype Batit as usual and make an egg and toast and leave Sean to sleep. I go to use the bottom bathroom and find bloody tissues strewn around. I knock on Sean's door and find him all beat up his eye is black and bandaged over the eyebrow and the back of his head is also bandaged. I asked what happened he says he doesn't remember but someone beat him up in a night club the security staff he thinks and he thinks the police took him to hospital and then brought him home. I send emails to his wife and mother.

Astrid shows up hostile with her father to take our car but decides to have a change of heart when I tell her about Sean. We do the letter, then while Sean sleeps, we go to the Embassy night club to find out what happened and recover Astrid's car which Sean left there. The Embassy was closed but Astrid talked to the night club manager on the phone and the manager said there was a group of people at a large table all ordering whiskey by the bottle.

When the bill was presented approximately USD 400 there was an argument about who owed what and a large Indonesian man started punching Sean and apparently also hit him with some club to the back of the head. The night club security took Sean to the hospital. We then went to the police station and they initially said there four people involved and they were in custody. We later learn that was for another incident at the same night club later and Sean's attacker was known but he would have to come give testimony to file charges. It would then go to court in a month. Well, I thought were not hanging around for more than a day unless Sean's condition is serious. Sean slept all Saturday I found some pills the hospital had sold him one for pain and swelling according to Astrid's interpretation of the label and the other antibiotics. I took care of Sean Saturday and Sunday feeding him fresh foods.

On Monday Sean's memory returned and he walked around and he could see but his eye ball was very red and still black purple under the eye. Astrid took him to an eye doctor and he said he would be alright no permanent damage. That night the cruising permit was granted but nobody told us until 7:00 p.m. and we had to leave by Tuesday morning at 6:00 a.m.

We got them to change the date to Wednesday the 7th at 6:00 a.m. Tuesday Sean and Tomas, Dion, Dodi and Ryan, and myself made final went to the boat and made preparations for the trip. I instructed Astrid to almost empty all bank accounts as we needed additional cash for the boat trip so she got Thomas to sign two checks leaving the amounts blank.

We loaded the house furniture, books and catalogs about the boat equipment and all of the nonperishable foods. I then learn without my authorization Tomas requested authorization again to leave by 10:00 a.m. and I was furious as no one asked me and I had a permit to get out and now maybe there was a huge delay and financial risk of changing it. We get back to the house, and Astrid, as an accountant, is more interested in how the final small amount of bills get paid, including six days of car use at $150 and a two hundred dollars of other minor expenses got paid. I told her to send budget to Vini as she has for six years and it will be paid. She was rudely adamant that they had to be paid tomorrow which was complete bull shit. She really pissed me off now lying to me it was evident after six years in our employment and paying her mother the cost of two new cars she did not trust us and thought we were leaving without paying USD 150 for her mother's car. I literally went crazy mad at her and Sean had to calm me down I was ready to start throwing the furniture if not at her through the window. I fired her on the spot and took all the money 53 million Rupia and told her I would pay the agent and the third-party company with the export license in cash the rest I would keep for the trip in case we needed money as an emergency. Why she had a brain fart and could not figure out how my office could send her money when they have been doing that for six years. It was clear it would go to her account as we needed Thomas to sign if it went to the PT OES account. The agent Edo, Melie, and a body guard came over that night with the cruising permit, and so we handed over 65 million Rupia about USD 5,000. He forgot the Customs documents so it was arranged he would bring to us in his boat the next day while we were underway. Sean would also come and take photos of the *Borneo Princess* under full sail.

Thomas had plotted a route for us to follow which would avoid shipping lanes, avoid coastal areas possibly occupied by pirates that would lead us south to the west of Sulawesi then we would turn right in the Java Sea heading west along the south coast of Borneo then turn right again in the South China Sea heading north of north east to Singapore.

6.0

Sailing the Sulawesi, Java and South China Seas

Wednesday 7th September we were off to the boat about 7:00 a.m. we would run out of tide at 840. We loaded the car with all the frozen meat consisting of three large 2-kilogram Snapper, five chickens, 1 kilogram of prawns (Shrimp) and two kilograms of hamburger meat. Also loaded were all the fruit and green perishables that we had been storing in our office house fridge for a few days. The Indonesia crew consisting of Dion and Dodi, Bill's brother in laws and Ryan had been with us since the beginning around six years ago. They all had brought their wives and small children as they would drive their husband's motor scooters back. They also got to come on the boat and have a look at about eight o'clock we recovered our stern anchor and then Sean untied our bow lines and we were off. There were a lot of boats and barges sitting nearby at Meranti shipyard and Tomas did a great job of maneuvering around them and out into the river in reverse. There was a strong wind blowing making conditions worse. We had not gone a mile when a large ferry boat pulled right out in front of us. Tomas slowed down and we had to follow them out of the river and into the bay at a speed of about 3 knots. Once out in more open water we hoisted the main sail and then the jib sail without a hitch. We were now sailing at 2 knots in an 8-knot following wind. Redo and Sean rang and they came up behind us and handed over our Customs documents and they took a great many photos. I had been waiting six years for this day but did not anticipate what was to happen later. We continued to motor sail out until our course change made sailing not feasible so we took in the sails. We experienced 4- to 5-foot seas with occasional 6-foot swells.

Everything was looking great we were all happy congratulating everyone on a good job with building the boat then at about 5:30 p.m. the port side engine started making strong shudders. Tomas immediately shut down the engine. Dion and Ryan went in the engine room and had found the flanged coupling that connects the propeller shaft to the engine had sheared some bolts and moved backwards an inch. We lowered the portside dive platform and I went in and had a look but it was almost dark. The Kort nozzle however was loose and was turned to port but should have been aimed straight. Tomas went in to confirm. Shit what to do. We continued a little at 3 knots on one engine a little further as Ryan and Dion repaired the propeller shaft problem but nothing for the rudder. We discussed the options 1) head back at 3 knots only 50 miles away 2) continue as is not likely and 3) find somewhere in shallower water to anchor then in daylight see what we can do fix the problem either fix the Kort nozzle, remove it all major tasks probably not possible or tie it straight. We continued for a few hours to shallow water anchoring area in about 35 meters water depth. We had 100 meters of anchor chain. Ryan did something that caused all of the anchor chain to just spill out fast but luckily it was tied to a rope that stopped us from losing it.

Everyone was tired it was 9:00 p.m., but the guys worked out a successful procedure of tying to the rope another rope and pulling the rope and chain up a few meters at a time using a long rope connected to one of the sailing winches.

Finally, we had the chain installed back in the chain winch and discovered there was a fitting you had to keep tightening on top of the winch to give it friction to the winch and keep it from freewheeling.

That night I made coconut chicken curry with fresh onions, garlic and sweet potatoes. Could not find fresh coriander leaves so used coriander powder and some strange tart lemony smell bunch of green leaves. Tomas will eat anything; the Indonesians would smell it, and if okay, slowly try a little.

Tomas woke me up at 5:00 a.m. am he seemed depressed and tired and wanted to go to bed and me to relieve him no problem we are on anchor. I sit on the computer and start writing this. Ryan and Dion wake up about six, and I make them a breakfast of eggs and rice with fresh orange pieces. We then decide to see if we can somehow fix or tie straight or remove the port Kort nozzle. I dive down and inspect it his hanging by a small piece of steel not yet broken 95% is broken away, Dion goes down and ties a line to it and they try to winch it up and break free. No luck. I go down and tie another line then pass it to the port side under the boat and get the guys to pull and release and try and break it. After only a few pulls it broke free. Dion installed the triple block assembly on the two stern davits and we transferred the lift point and hoisted it on deck. Tomas woke up and came out a little shocked and in disbelief of what we had done successfully removed a quarter tonne steel cylinder and got it on deck. I was hoping to pull up the anchor too before he got up and possibly get underway but he was too quick. I made him the same breakfast.

At first the starboard engine would not crank so we suspected a battery problem and tried other batteries to no avail. Finally, Ryan finds out it is a faulty Chinese isolator switch, shit from China, and so we remove it. Everything was going good for an hour or so and we were underway when I went to make him some tea to discover my room floor and hall way underwater by about a half inch. I discover I had closed one of my portholes the previous night but did not lock it.

While underway water poured in filled the floor my bed and you could use all the drawers as a fish tank. The team of Ryan, Dion, Dodi and myself got to work with buckets, mops and squeegees and cleaned it up. My new queen bed mattress soaked so they took it on back deck to hopefully dry out one day but I knew it was ruined,

I sat at the wheel and navigated a while and talked with Thomas, then made him lunch left over from chicken coconut curry with hot fresh red chili peppers and he loved it. About 2:00 p.m., we hear a similar noisy rumbling like we did the previous day. Tomas shuts the engines down. Ryan and Dion check the engine rooms—no problems. Tomas suspects it's the starboard side Kort nozzle. We both dive, however we have 6- to 7-foot swells coming side on to us making our diving difficult and dangerous. We confirm it is the same problem the Kort nozzle torque arm has sheared and worked loose but not really broken enough to just break it off.

We try to remove the four nuts holding the bolts but is hard to get at especially when you are being knocked around by the waves hitting your head against the dive platform and Kort nozzle and scratching yourself on the sharp barnacles.

We try and loosen many times and they turn and so do the bolts on top. We try all sorts of ways to pull and rock and break it, but progress is frustratingly slow. Finally, we decide to throw an anchor out, tie it to the Kort nozzle, and go forward hoping this will break it off. While they were preparing the anchor, I went down again and found one nut had worked itself off and the other three were finger tight so I easily removed them and got out of the way. Before I returned to the surface it fell away and he were happy. The guys successfully recovered it on deck and after only two hours we were underway again with no rudders. Tomas advised we were going about a knot faster now and steering was not a problem in fact it was easier we would have to adjust engine speeds to turn. I went in to wash all my cuts and scrapes. I was glad I had trained as a construction diver which made the job possible.

We continued all night and the next day through 10-foot swells some head on others side on. The boat and especially mast and rigging held up well. The boat movement was mainly pitching in the bow with large splashes of water hitting the Saloon underside throwing water to the wood panel trampoline witch also held up well no discernible movement or breaking of wood like before.

That day a fisherman or pirate in a small outboard tried to cut in front of us twice and stop us and we almost hit him. Tomas kept going I suggested we go back and run the idiot over and sink him.

I do a night watch at the helm on the bridge from 10:00 p.m. to 2:00 a.m. and navigate trying to stay on our route and miss other shipping traffic. Tomas relieves me at two and I go to sleep, relieving him at 6:00 a.m. to 10:00 a.m. and so on.

It is Friday Sept 9, 2016 morning 10:00 a.m. everything was good we continue without any major problems. The starboard generator failed so we switched over to the port side generator. That night the starboard engine sounded alarms. I switched it off as the guys looked at it. The fuel filter was cleaned and we were off. During the shutdown, about an hour, I successfully navigated around what looked like an oil and gas loading jetty all lit up. With the strong wind 14 knots from our port side, it would propel the boat and twist clock wise. I would maintain heading by running the port engine at 600 rpm and also achieve an 8-knot speed.

Sean rang Dion as Dion sent him photos of our steering systems on back deck. Sean thought we were in trouble and how would we get back. I replied we are going to Batam as planned then the phone went dead. He was actually south of us In Surabaya visiting with his wife Yusi her mother and another wedding celebration a Muslim one. My son and brother and boat captain Thomas ex Nuclear Sub weapons officer were now Muslims. Who would have thought?

We had seen 10 ft swells and went through them against them and side on to them. The boat would mainly pitch into the seas and bang under the saloon roof but its roll was reduced to less than 5 degrees which as not only manageable but you did not really have to tie anything down Even Astrid's office swivel chair did not roll

That night I made fresh spaghetti sauce with spaghetti and meat balls using only fresh onions, garlic, tomatoes, green pepper and mushrooms. It was a lot of work and I lost a lot of sweat in the 90-degree galley heat as you could not open any port holes underway. The port generator shut down but Ryan and Dion and Doddy were able to quickly fix it only a fuel filter problem again. I sliced open a large watermelon for everyone.

Saturday morning Sept. 10, 2016 little did I know that in exactly one year from that day we would be hit by a category 5 hurricane (cyclone) with 180 mph winds named Irma. I took over at 3:00 a.m. and could not keep my eyes open. I was sick with dysentery and at 6:00 a.m. got Ryan to take over the wheel actually we had no wheel no joy stick and no automatic pilot. I went to the toilet for what seemed an hour showered and slept until 8:00 a.m. when Tomas came to relieve us.

Saturday was a good day no shutdowns we kept moving I pushed the boat to 9 knots with jib sail out especially when a suspicious fisherman boat kept coming towards us. I half suspected pirates, but we figured out that we could probably outrun them, or if we had to, we could easily run them over and sink them.

The guys had fixed the other generator again the fuel line was clogged. They tried to fix my bathroom grey and black water pumps but could not; they were fried. Later Dion found a second spare grey water pump and installed it. I could continue to take a shower and pee down the drain but all shitting had to from the port side head as we had no way of emptying that tank and if it overflowed what a smell everyone would be sick.

We had another run in with fisherman that day maybe they were just curious but we did not want to take any chances. We were in a very isolated area only remote villages on the large island of Borneo to the north and 400 miles south the long Island of Java.

For lunch my leftover spaghetti sauce, some red and black beans thrown in and the rest of the hamburger meat already thawed out and the previous day's leftover rice and throw in some fresh cut chilly and that was lunch, and everyone ate a sort of red wet rice with meat and beans and tomato sauce. It was good chili con carne with rice, I think it is called.

I was glad things quit failing but know we had worked the bugs out with all the engines and generators. The props were not even a problem as I suspected earlier. Whatever speed you ran the engines they would deliver flow even at 1700 rpm. We could not get more speed without tinkering in the engine room with the throttle cables.

That day I inspected all the welds of critical structure welds that could fail in the front 500 deep beam connecting the port and starboard pontoons and the reinforcement doubler plates for the mast side stays. No cracks actually did not expect any, boat structure was designed well and Davidi were good welders and proud of their work and the OES team also would have inspected. The Kort nozzles should have had doubler plates and this was missed by all. I did not do the structure design of those Davidi designed what the local boat industry was supposedly using.

That day Tomas was asleep when I get four boats coming towards us from the west and we are heading west and they were heading eastward. Thomas had installed an AIS system which communicates with all the other boats in the area radar and broadcasts the size, registration, speed and bearing of the vessel. I was worried especially not really having any good steering capacity but did my best went to the right of the first one and second one and to the left of the third one. Thomas showed me how to operate the system before but this time I was responsible.

We ate left over chili con carne and the remains of the chicken coconut curry with a fresh salad.

I relieved Tomas at the helm at midnight until 6:00 a.m. Sunday. The moon was shining bright and always was in the west that night so he worked out a more effective way of steering aligning the moon with parts of the boat. He would keep the starboard engine running at a good speed of 1350 rpm and power the port engine twisting the boat clockwise and it would naturally twist counter clockwise from the wind and wave conditions. You would allow a certain swing of the bow to the left and then power the port engine until the moon shine off the water shifted to the right where you wanted it. We had USD 50,000 of instruments and navigational devices including a steering wheel, toggle switch, automatic pilot system, three Garmin GPS track plotters and a Simrad positioning system and we are navigating by the moon and stars which was easier than looking at a compass which puts you to sleep as long as there were no clouds.

Well, the clouds came out so I had a system power the port engine to a heading of 286 degrees let off the port throttle and let boat twist naturally back to 266 degrees before powering again. This kept us sort of straight with a variation of 10 degrees off both sides and was easy to execute a simple procedure that worked well. I was navigating by moon light and then my system when about 20 fishing boats all lit up appeared over the horizon. I deviated off our plotted route to the left about a mile to avoid them. The boats were everywhere constantly appearing over the horizon as bright spots you could see 10 miles away or more.

On the Simrad AIS I see a large freighter heading for us from the west on our starboard side meaning he has right of way but the accuracy or distance is not easy to judge you need to actually see them to make it easier to avoid. The freighter has no navigation lights so I cannot spot him. According to the AIS screen simulation we are going to collide and soon. Then I spot him a quarter mile away death coming past us with no warning lights. I wonder how many Indonesian fishermen that don't have all these collision avoidance systems disappear each month.

At about 5:00 a.m. I get another AIS warning this time a 450 ft long oil tanker coming from behind us and to port. We have the right of way this time but I don't want to be floating in the Java Sea dead or alive right or wrong. I notice on the AIS screen the tanker keeps correcting course to avoid a collision I presume. I can't imagine what he is saying probably "what is with these *Borneo Princess* private yacht fuckers they keep changing course from 266 degrees to 286 degrees." There are fifteen ships in that area mostly lit up fishing boats I think they use lights to attract the fish, crabs or octopus, I am really shitting myself I can't see this tanker and we are on a collision course again the vessel has no navigational lights. I wake up Tomas at about 5:30 a.m. It's just starting to get light. I need his help. He found it a green old rusty steel death ship. I continued on our course just to see what would happen. It passed and cut in front of us to the right about 200 yards away. Not my day to die.

Sunday was sort of a strange peaceful day hard to explain. I made Tomas and I some very tasty and healthy granola cereal. The skies were blue and everybody was relaxed and not stressed and there were cool sea breezes. Dion had adjusted the engine throttle cables to get more rpm, and I told Tomas, "Let's open up this baby." We got 1700 rpm and a 10-knot speed and were consuming 11 gallons of fuel per hour close to my original calculations making me feel a little confident in my potentially new career as a marine engineer. Best of all no noisy or excessive vibrations meaning I was wrong before we don't need to go to dry dock and have engines aligned or propellers balanced. Tomas and I even believed we could rebuild the existing Kort nozzles in Batam, but his priority was to get boat to Singapore ASAP not knowing what type of bureaucratic delay pay me money crap we may face there.

Sean would be in the area having other company jobs like fixing and selling our submersible excavators I bought for a song in our corporate expansion spending spree of 2010 where we completed a job in Angola netted 1.8 million and spent it all on investments in a new

Indonesian ship yard and new method of trenching shallow water pipe-line shore approaches including a 100-tonne crane and 220 feet heavy capacity cargo barge. We did get this boat out of all the efforts and

the crane and barge were for sale for 60% off. The Angola job so the culmination of twenty-five years of me trying to build a pipeline rock trenching and acquiring equipment and $500,000 hydraulic hoses all along the way paid for by clients on other jobs. They paid about 25% of the cost to build it during the mobilization fee but never used its rock cutting capabilities.

I did my nails I mean clipped my nails and washed my clothes in the galley washer dryer before coming on shift at noon. I may have put too many clothes and sheets as it washed but would not dry and we spent four hours trying to figure out to work the damn thing. Reminds me of the day it would take four hours to try and prerecord a TV movie on a VCR. Tomas relived me at 5:00 p.m. and I went down to prepare Sunday dinner consisting of fried snapper using Louisiana seasoned flower given to us by red neck Stacy, fish head soup and fried rice with carrots, onions and green beans.

Everyone loved it especially the Indonesians wanting me to fry up the chicken with the same flower. Everyone used my nail clippers that day.

When I went down to cook the kitchen felt like a sauna about 120 deg F. The clothes dryer had been trying to dry too many clothes for a few hours plus you would not normally open port holes while underway especially after the last fiasco where we took in a tonne of water and shorted out the shower drain and shit tank pumps. Anyway, I brought in my room fan and had to open the two port holes it appeared safe. An hour later swash about 20 gallons of wave water comes in one. Dion and Ryan had buckets with dust pans and squeegees and mops and also rinsed the floor with fresh water and so within fifteen minutes we are ok but kept port holes closed.

At about 6:00 p.m. after dinner I went to check on Tomas holy Christ it had just gotten dark and we had over fifty all lit up fishing boats in front of us and on the sides of us. I don't know why but felt we were the offensive football team him the quarter back and me the full back and we did a quarter back sneak rights up the middle of them. You could imagine the stares and the foul words when this large sleek white ghost appears and passes between them.

Next morning Monday at 12:00 a.m. I come on duty things were pretty quiet no more fishing boats but you had to keep an eye out on commercial shipping traffic. These captains are like the 16-wheeler truck drivers that think they own the roads but there are no phone numbers painted on the side of the ship to ring if you have a complaint. I am positive they must plow through many small wooden fishing boats every year. Tomas showed me on the charts there were two large wrecks, can't imagine why, along our route and were buoyed with flashing lights. The first ship was about 24 miles the second about 40 miles up from us. Of course, if navigation around the submerged objects wasn't enough five container or oil tanker ships approach.

I got up for noon shift and more of same with tankers. After I prepared a kilogram of prawns cooked with Stacey's Louisiana Crab Boil plus some green vegetable and papaya which everyone loved. I stepped on a piece of glass in galley that cut into my foot wiping blood up with my foot while still preparing the meal. Later cleaned it with some alcohol I found in Bill's drawer at the office and put on some Neosporin, a great antibiotic.

On Tuesday's early-morning shift, it started to blow hard from astern, 18 knots at times. Swells came up from behind us and started to spill a little water into the stern deck drains. I had Ryan plug them. I was worried that water may spill over the transom or blow into the engine room, so I had the port engine room deck hatch closed and fans switched on. They worked, but there were butterfly valves or dampers that appeared to be closed. Dion informed me they were supposed to be open when the handle was at right angles unlike any quarter-turn valve I had ever seen.

I got the idea if we could outrun the waves, they could not spill over the stern. I was wrong going; fast made it worse. As the boat would go up, a wave then transom would now be lower. Better to go at the same speed as the wave.

I told what I discovered to Thomas and he laughed, already knowing this. In the end you went a knot slower but did not have to constantly work engines at high speed and powering up a wave using much more fuel.

I made ham burgers chicken fried steak way by coating them in the Louisiana spicy flower with lots of "Slap Your Momma Powder" and frying them in oil. Indonesians all gave me thumbs-up while eating.

Wednesday, I started my watch at 3:00 a.m. to 9:00 a.m. today putting Tomas on watch during entry into Batam and the careful navigation required and docking at the marina next to all the other million-dollar yachts. Everything was going well then "Fuck me dead" there were no less than thirty-five ships coming straight ahead at me. I noticed to the left was shallower water and I could head there I only drew 1.5 meters and the ships about 20 meters so they would not go there. Problem was there was one ship coming up behind me on my port side but about five miles back so as my only choice to steer left and cut him off and he would have no other choice but steer right to also avoid the shallows. It put both engines at full speed and we got 10 knots for about an hour and it worked I got safely in front and crossed his path and he turned right to avoid us and the shallows.

I felt like a submarine commander in a World War II movie making decisions on what you expected the opposition to do. As I cut across the shallow water, I also knocked about fifteen miles off our trip as I had taken a short cut across the right-angle bend Tomas had plotted. There was an exposed wreck on the charts I also safely avoided. Of course, now rain squalls appear and we have to close up the wheel house windows and my visibility is greatly restricted even with the wiper on but at least no other idiot would enter these shallows I hoped. Showers stop and sun comes up and I can see chart screen the night screen is still on and I need to switch the much brighter day screen on. I fumble for twenty minutes no good and worry I might delete the route so finally have to wake up Tomas early.

I also had noticed during the night while peering into the port engine room a larger spray than usual coming past the shaft packing seals. The *Borneo Princess* was no longer a princess. She was a Queen. She had two shaft holes like all women and this one obviously her vagina had been opened up and she was gushing water no longer a virgin. She had been pounded up to 500 times per minute for seven days by an American named Caterpillar with a long shaft from a Tierra. Her anus hole or starboard side also pounded but not leaking much maybe re alignment of port drive train required.

7.0

Arrival in Batam

We finally arrive and pull into a sea wall protected marina area with large a large Mediterranean-style red tile roof villa complex built into the hill side and a deep clear water marina with many beautiful expensive pleasure yachts. The *Princess* could now strut herself here and hob knob with other princes and princesses. Tomas calls the manager and there were five people to help us tie up. Tomas carefully backed up into the 14-meter-wide berth and we tied off in the center and pulled the boat back to disembark from the stern using Joe's ladder.

We made it some 1100 nautical miles in 6.5 days averaging about 7 knots. Fuel consumption about 6.5 gallons per hour and used about 2,028 gallons or about 7,635 litres leaving a balance of 4,375 litres of the original 12,000 litres we started with.

I noticed one of our guys had pumped the engine room bilges and there was oil all over the water of this otherwise pristine environment. I wondered when we would be asked to leave.

8.0

Singapore

Most of this section and chapter, since I was not there, has been completed by a review of emails with Tomas, and Sean.

Tomas after two weeks in Batam obtained a building Certificate from Davidi and then successfully cleared the *Borneo Princess* out of Batam, Indonesia into Singapore.

He also found a place to berth for a month on the far side of Singapore. We did not trust their agent whom wanted to be paid direct two-month rent one as security deposit in advance and feared the money would never make it to the marina owner so we paid the Marina direct. We only stayed a month and so they owed us a month's deposit which we had to fight for and threaten to go wreck the place if not paid back promptly.

Sean, Tomas, Dion and Dodi took down the mast head, lights and radar by hand while at the ship yard. They did not stay on the boat they stayed at a hotel on the other side of the island closer to the port.

Thy had to put up with a lot of expensive bureaucracy with the Singapore Port Authority requiring an expensive tow by authorities into the Port, excessive boat insurance for which we had none and had to cover removal and salvage of boat if it sank in the port. We actually co-pied a Loyds quote put Lincoln Mutual Insurance on the letter head and this was accepted and *whamo*, I am also in the insurance business.

We also had a problem that the Singapore Port authority tried to charge us an additional USD 25,000 for their cargo costs as they meas-ured the cargo volume from the top of the mast these fucking Chinese cheating bastards. At the time I did not trust United Shipping so told them I was going to cancel the shipment which they could not afford as they already paid for the ship and so they reluctantly agreed to dis-mantle the mast and store in the holds of the ship at their cost to avoid these additional costs.

9.0

*Borneo Princess arrives
in Port Everglades, Fort Lauderdale*

We received advance notices of the anticipated arrival of the *Borneo Princess* and the thirteen containers and I learned I had to organize everything as this was not a normal container shipment off a container ship. I spoke to United Transport and they initially agreed to offload boat off the super tanker deck and set the mast and boom which was removed by them in Singapore. They would also unload my thirteen containers. They had subsequently changed their minds and demanded USD 5,000 to unload and install the mast and boom otherwise they would just place on my deck. I reluctantly agreed feeling screwed.

I arranged weeks in advance for a clearing agent named Heidi Marks my mentor, and probably one of the luckiest things I have ever done, to submit all the necessary paperwork to customs whom would then inspect and approve or not approve and could levy large duties for the USD 1.3 million ship under declared at USD 250,000 and the thirteen containers holding hundreds of thousands of dollars of cargo in the form of my personal tools a large hydraulic power pack, 1600 HP jet pumps, half a mile of 2 inch 5000 psi hydraulic hose on a large hydraulic powered hose reel valued at USD 500,000 alone. In addition, all the salvaged items off the Tierra I bought from Sam except the engines but including the V drives was also shipped.

My customs duty and sales tax could have been hundreds of thousands of dollars but I avoided it months before by getting the ship flagged Jamaican and therefore imported as a temporary import. On this basis it was limited for commercial use within the USA provided I pay customs and sales tax and obtain an exemption to the "Jones Act" called a Myrad Agreement which I subsequently got and could not take more than twelve paying passengers. I could however use the vessel commercially between US ports and foreign ports with no further fees or exemptions which is what I had planned doing international dive charters. What I did need however was a cruising permit to take the vessel from Fort Lauderdale to Key Largo but neither my agent, my Coast Guard, or Jamaican registry agent or United Transport advised me of this.

The containers were cleared as the personal tools of a retired old American guy that came home after 35 years abroad all organized bought and sold by customs by the lovely agent named Heidi Marks. The fact the boat arrived on New Year's Eve may have also had an effect as no custom agent was there to inspect the containers or the *Borneo Princess*.

I had prearranged for the Port of Port Everglades to load my containers sitting on the edge of the wharf and transfer to temporary storage. I would arrange pick up and transport later the following week. They would load onto my trucks, and then I would transport and unload at my five-acre Redlands estate.

I was not really sure when the ship would arrive at the wharf no one did or what exactly the procedure was. I knew I could not leave it alongside the ship and was not aware of the wharf conditions and nothing was communicated clearly from United Transport. Luckily Tomas Simonson my captain had arrived on December 30th and I picked him up at the airport in Miami.

During that pick up, I accidently backed up into this old bomb of a car doing little if any damage at which point the passenger, a thirty-year-old muscle-bound hothead ran out screaming. I screamed back as he took off his belt and ran to my widow to fight. I got out slowly kept my cool thinking I am fucked I don't even have a knife to defend myself against this cage fighter looking asshole. I finally quit talking to him and spoke to the black girl driving the car whom remained calm and I gave her my contact details and showed her my driving license and insurance papers. The big guy looking cheated he did not get to kill someone that day. I should have waited for the police I am sure he was wanted for something and probably her too. We drove home to Key Largo me still feeling a little shaky getting lectured by my wife the whole trip how I am old and can't get into fights anymore especially with young strong men. I never in my life before backed down from a fight but that night I did I feared for my life.

Next day New Year's Eve Tomas and I drove to the Redlands to see Nicki whom knew boats and wanted to come with us bringing the boat to Key Largo. I had met Nikki, a sixty-year-old Venezuela man with a half American Indian wife named Susan whom also owned and lived on a five-acre farm on SW 216 St a few miles down the same street my property was on in the Redlands. I met him one day when I saw a sign "Tilapia for Sale." I had bought some Tilapia in the previous months for my large quarter acre lake I had dug. They were to keep the mosquitos in check and also eat any algae that may develop. Susan had a prestige job was a graduate degreed person and worked directly for the Surgeon General and was also doing welfare work for the Indian tribes in Florida. Nikki and Susan had been to the house a few times and they also had a house in Key Largo they shared with Carlos, a Cuban Lawyer who will figure prominently later on. They had met in Oklahoma when Nikki was a cattle farmer and now they raised plants fish and fresh water lobsters called maron.

While at Nikki's we get a call come now to the ship we will be unloading your boat soon and you need to be here. We all got ready went home to Key Largo grabbed some clothes and Susan was to make some sandwiches and we came back picked up Nicki and were off. The instructions we got from United Transport were vague and it took along while to find the ship. Once we got there, we were shocked to see they had put the *Borneo Princess* in the water without us to check and tighten the shaft seals or making other preparations.

The ship was massive about 700 ft long 100 ft wide and the deck 50 ft above us. We boarded the large ship and our *Borneo Princess* looking rather small in comparison was sitting on the starboard side and heaving and rolling from the large swells that somehow made it into the port from the severe offshore weather and reverberated off wharfs like sound waves and found us. Tomas was able to climb down a rope ladder and jump 10 ft down and 5 ft out onto the deck of the *Borneo Princess*. Nicki or myself knew we could not do that without breaking a leg or a knee and so felt bad for Tomas whom looked sad by himself pulling out heavy turnbuckle rigging for securing the nine mast stays.

United Yacht was getting ready to set the mast so we thought but all they did was drop it in its slot and connect the boom then their crew left. They gave us one guy to help after me complaining to the management on the phone and also asking for transport to the *Princess* as we could not climb down. We drove to another location and a small transport boat took us to our boat after first trying to con us out of money by arranging a tow to a fuel dock for USD 1,000 and then dockage for USD 1,000 per day. He advised that there were 40-knot winds blowing outside and no one was going out through the heads. I got his number and said I would call him but never did.

We worked all through the night Tomas, Nikki, myself and one United Transport worker securing the six side stays the two stern ones and the jib sail/ front stay. The boat people wanted us away as they had to get underway and their ship was costing USD 2,000 per hour.

Connections were made more difficult by the swell and slight rocking of our boat from the long swells. I had to finally lie down for an hour or so no more energy. Tomas got the boat ready to get underway. The Port Side engine would not start and luckily Nikki got it running by connecting a battery charger to one of the batteries.

10.0

Off to Key Largo

We were off. I discussed the options with Tomas and we agreed to head for Key Largo get the hell away from here. Problem we had is United Transport could not install our mast head with all our running lights and one of the navigation lights our red port side light was not working. All we had was a green starboard light and our white stern light as we went out the narrow channel heads in complete darkness and there were large ships probably 700-ft long coming in. One ship captain got excited and kept blasting on the VHF radio "Small ship dead ahead your Port to our Port side do you understand Our reply "yes sir port to port" as we went to starboard getting close to the large rock groin Americans call a jetty. At one point a port pilot came along side to see if we were ok drunk or what. He shouted stay to starboard and we gave him the thumbs up as this massive passenger liner slipped by with about a thousand people looking down at us over the hand rails from various levels. The crew looked in amazement that such a little boat without lights would head out into the severe weather.

Hell with it. I knew I could dive in and swim if I had to. I was never worried. We headed out in large twenty-foot seas at the cut entrance breaking over our bow. Luckily the guys in Singapore did a good job securing the wooden trampoline panels on the bow. We made a hard right and headed due south down the coast next stop Key Largo. The ten-foot swells tended to rock the boat but not much we stayed in shallow water about fifty foot outside the reefs but inside the heavy seas. Nikki and I adjusted the turnbuckles now and then when the lines got loose and whipped around.

We watched the sun come up to our left and felt elated somewhere south of Fort Lauderdale. We continued all day watching the Miami Beach Skyline appear and disappear on our right with all the large hotels and apartment buildings lighting up with the sun shine.

Close to Fowi Rocks light tower marking the entrance through the reef to Key Biscayne, I remember seeing a Coast Guard Cutter behind us and half expected it to board us. At the time however I was broadcasting and advised traffic on the radio who we were and that we had little steering and to give way. The Coast Guard must have decided we were not smugglers and turned back.

Around 6:00 p.m. close to dark we made it to South Cut Key Largo. It was not yet high tide but tide was still rising so we decided to chance entry. We ran aground no big deal with the strong steel keel and recessed propellers and rudders above the keel. We eventually after an hour or so were able to break free in reverse thanks to the rising tide. I rang my friend Sam Stoia; he advised us to go and anchor at Rodriguez Key, which, although about a half mile offshore, was protected from winds from the South and West and the reefs to the east would break up any large waves so we headed there only about four miles south of where we were in twelve feet of water.

After tightening all the stern tube gland seals and putting only one 160-pound grab or Bruce anchor out following my captain's advice, we called Tow Boat USA and they came and picked us up and took us home to my house on Largo Sound. The charge was USD 200. We would return the next day.

11.0

Blown on to Rodriguez Key

When we Tomas, myself and Batit, my wife, arrived the next day in the afternoon from by 28-foot Tee Craft and were shocked to see the *Princess* had been blown to the South East and ended up grounding on the South side of Rodriguez Key within the Marine Sanctuary boundary.

I checked the water depths and were grounded on the starboard stern in just shallower than 6 feet of water. The *Princess* draught was 6 feet and luckily again the tide was rising. I took the *Princess* Port Side bow anchor in to the wind which was blowing 25 knots and pulled with little effect than placed it hoping we could pull clear with the anchor winch.

I then unfortunately got on the radio and called "May Day May Day" which was an emergency distress call. A Coast Guard helicopter was there within thirty minutes and directed a local private boat that came by. There was a lot of wind blowing us further inshore to the island and the 40 ft cabin cruiser with inexperienced crew were reluctant to come close to us and it took a while but we did manage to give them a line connected to our port stern and asked them to pull towards port against the wind.

A few minutes later Sea Tow a marine towing company came out and the operator advised he should pull from our port bow in to the wind away from Rodriguez and we agreed. Eventually we moved off and dropped our port anchor with 300 ft of chain on the sea bed. We then recovered onboard our starboard anchor that had been dragging and Sea Tow towed us back to where we were this time in a little shallower water in about eight feet of water and with a lot more anchor chain out.

It was now dark and we were tired and had been up all night the night before. Tomas tied the *Borneo Princess* stern on to the waves and me not thinking about it too concerned with what had happened. After Sea Tow got us to sign the necessary papers, not knowing we were being set up for a salvage claim, they came back to say the Tee Craft was sinking in the stern. The stern transom was under water so I thought we had lost it.

The engines were also partly submerged so I immediately started them. I reacted quickly put it on the *Princess* stern and lifted the Tee Craft stern above water with the stern davits as Tomas pumped it out with the gasoline bilge pump he finally found and got started. I went into the cabin and everything was two feet underwater including the batteries. Batit got in and started bailing out water with a flat faced bucket. We saved it and the engines and even batteries were fine.

Later Tomas and I in the dark and strong winds unsuccessfully attempted to recover the *Princess* anchor we abandoned. We gave up picking up Batit and headed home.

The next day we recovered our anchor off the sea bed and reinstalled it on the port bow. This time I insisted we have not only two anchors out to the ENE and ESE and so had purchased a second hand 80 pounds Danforth anchor and placed it from the port stern to the North.

One day we got a letter from Sea Tow demanding USD 2500 or they would claim salvage rights on the *Borneo Princess*. The agreed amount was previously USD 400 and I wrote to them accusing them of extortion and considered going to court. Their case was weak as the boat was already floating when they arrived. In the end I did not want any adverse publicity or possible fines as we had drifted and grounded in an environmentally sensitive area and just paid them.

12.0

Rodriguez Key First Time

One day, a young blonde girl came out in a boat looking for a job, and let us knows her sail boat anchored nearby had blown away two days before from a freak westward wind the same night we broke anchor and ended up on Rodriquez Key.

We would continue to come out and find the boat had drifted around a little and the Danforth anchor did very little. We would untangle anchors and reset. We noticed our stern light was out and the house batteries were flat and should have been charged from the solar panels. We discovered our port side bilge pump was stuck in the running mode and was using all the battery power. We fixed the pump. On one trip we took Carl French, a good friend an experienced but now retired international Yacht broker I met through Steve Levine a neighbor friend, out to the *Princess*, and a small fire broke out from the starboard bilge pump. We quickly extinguished it although way down in the wet steel bilge the fire could of have done little damage. We later replaced both pumps, and since very small amounts of water were entering the boat through the gland packing, we decided to switch all bilge pumps off except when we came out and were there.

Routine trips to the boat would involve starting the two generators, charging the house batteries which ran the bilge pumps and navigation lights, checking and repositioning all anchors, pumping out the bilge if needed and starting the engines. In effect daily trips were about keeping the boat in one place, mechanical systems running and from sinking.

A day before Tomas had to part for Indonesia to get his wife we went to the *Princess* and we were going to routinely start the Port side engine when I checked the oil and found it was milky white and therefore had water in it. Before he left, we decided to pump out all the contaminated oil. Nicki was asked to help and as Tomas pumped for some reason Nicki pulled the hose out of the bucket spraying Tomas in the chest with milky oil. Tomas asked me to not allow Nicki on the boat after that. I arranged for Caterpillar to come out and inspect the engine and advise us what to do.

Sean came out from Australia a few days after Tomas left and we went over and over every aspect of the boat operation to see how water could possibly get in. At first, we believed it was backing up through the wet exhaust large diameter piping in which the bilge pumps discharged into. Later that proved not to be as I could put my hand in the discharge outside underwater and there was a large air gap indicating no water was backing up. Later Sean found the cause. There was a small one-inch ball valve from the gear oil cooling water that discharged into the overboard dump system for the generator.

The problem was if you ran only the generator and this valve was left open the water could back up into the engine. The solution was to keep this vale closed when not running the main engine and in addition to add a non-return valve for double protection.

Caterpillar came pulled off the heads flushed the engine with diesel fuel several times drained it and refilled with new oil. It started up no visual signs of rusting or corrosion or apparent damage. We were lucky we discovered the water before we ran the engine.

Sean and I got wind of a huge storm coming that night with westward winds, so we went out and bought a second-hand stainless steel 150 pound plow anchor. We went to the *Princess* that afternoon and placed it directly astern pulling it tight with our stern sailing winch. We also repositioned our two bow anchors and Danforth breast anchor. We decided to spend the night as winds of up to 50 knots were forecast. The winds blew about 50 knots from the South East and for most of the night and they did not affect us very much. We did move according to the GPS about 200 ft to the west as the grab anchors dragged. You could sleep in the front cabin suites and not even know the wind was blowing. At about 4:00 a.m., they came around from the west and this caused the Tee Craft which was tied to the stern to swing around and was port side on to our portside banging into the boat. We heard the banging and went out and moved the Tee Craft and tied it to the bow so it could swing harmlessly to the East. There was some bumper rail damage. We then went back to sleep.

The wind blew 40 knots from the west for fourteen hours and the *Princess* did not move. The plow anchor dug in very deep as we would find out later and the 100 feet of chain we put out and 40 feet of rope line were almost horizontal leaving the stern of the *Princess*.

Man, what an anchor; I was learning all the time. Satisfied the boat was safe for almost any wind that could come or way we got in the Tee Craft and headed to Catamaran Boats owned by Sam Stoia where we had been keeping the boat to make the USD 135 per hour trips by Caterpillar and our fuel costs less.

It was hard to get out and off their wharf if you had a strong east wind causing me to damage one propeller off the rock bottom and we were also limited to the tide at very low tide we could not easily get out and had to lift our engines for over a half mile not having much control in strong winds.

13.0

The Story of Sam Stoia

Sam was a very long-time friend dating back to 1969 when I met him through my dive buddy Rick Driver. Sam and I used to go deep tank diving and spear fishing together during the early '70s. In fact the ship yard he now owned was owned by a family called Lowe and I first took Sam there where we rented boats.

Sam found me in 2005 on Classmates.com and I came out and visited him and started coming out and going diving and looking for old Spanish Wrecks, Spear fishing and just having a good time every year with my son Sean and some of my offshore crew. Sam would donate the Tee Craft as his contribution to the treasure diving expeditions and I would purchase later in 2009. I would in 2010 buy my current house on Largo Sound a place to berth my *Princess* whenever it was finished through the efforts of Sam and his wife Alana.

Sam also put in my wooden piles in 2013 for a good fee to berth the *Princess* and I had bought all the components of a fire-razed 50-foot Tierra in 2011 including two 1000 HP Caterpillar Engines used for running my big pumps and the shafts and propellers used for the *Princess*.

Now in 2017 Sam loaned me two new 13 ft. x 13 ft. 7-ton lift bags he had gotten very cheaply which I needed to lift the *Princess's* stern so I could get into Largo Sound. With the bags the maximum draft was 5 feet and Largo Sound had between 5.5 ft. and 6 ft. of water depending upon where you were and extent of the tide. Sam also allowed us free of charge to berth the 28-foot Tierra at his shipyard which made trips out to the *Borneo Princess* less time and fuel expensive. If the wind was blowing hard from the south, it was difficult to get away from the dock causing me to chip my port propeller.

Sam had introduced me to Johnnie Rodriguez, Ramon and Mike his cronies. They gave me good deals carrying heavy furniture upstairs including a baby grand piano we put upstairs over the back veranda hand rails with a rented fork lift. Johnnie also helped me cut trees at my Redlands house and unload semi-trucks carrying five sets of large CNG bottles.

Sean and I installed and inflated the lift bags without a hitch. We got the stern up a foot and now the entire vessel draught was about five feet so believed at high tide we could get behind the house. Sean had to return soon and lucky for me a guy named Johnnie Rodriguez, although his real name was Gustavo Barera, called me looking for work. Johnnie after two months had just made bail and gotten out of jail and was in bad need of work and money. He was jailed for not reporting his new address after he moved to Las Vegas to live with his brother and work. You see Johnnie was listed as a pedophile and spent ten years in prison sentenced by an Afro American black judge whom apparently was jealous of all the government grants and subsidies given to Cuban refugees. Johnnie's only crime was after he came to this country as a refugee, he was arrested for living with a minor, a seventeen-year-old girl who he loved. The girl's mother a crackhead who wanted money from Johnnie had him arrested.

Sam and his wife Alana started going weird once the *Borneo Princess* arrived. They seemed jealous of my boat, and Sam started getting rude to me. On one occasion after I paid him to pull my 28-foot Tee Craft boat out of the water and also was paying a daily storage rate while I was doing some repairs and under my boat painting, he drove by in his car and insulted me called me dirty for having paint on my hands and he also insulted Joe my welder whom came by to discuss things about smoking cigarettes on his property. I needed to get my boat back in the water to pick up Johnnie as the *Borneo Princess* was coming back to the Keys from Miami. Even though I had paid Sam and his wife over USD 120,000 in services over the years, they made it clear and in Alana's words "No Cash, No Splash" meaning they were not going to trust me for money. I had to pay the USD 300 before they would launch my boat. On one occasion, Alana drove to my house unannounced to see the *Borneo Princess* and walked into my yard without the slightest bit of courtesy to advise she was there to see the boat. They then demanded the bags back they had loaned me. It took me about 8 hours to deflate clean fold and return bags.

14.0

To Largo Sound and The Tomas and Sella Story

Tomas had now returned with his girlfriend Sella, a nineteen-year-old Indonesian Muslim girl. I had taken them to get a marriage license and allowed them to live upstairs in our second bedroom until the *Princess* was at the dock and then they were to live on the *Princess*. I also took them to see many apartments for rent.

Sella did not like America and the Florida Keys it was too expensive compared to Indonesia and she did not like living in our house either. She was very insecure and excessively controlling of Tomas and I knew it, Tomas and Sean apparently did not understand this. Things were made worse in that Sean would sleep in their room as the bottom apartment was rented out. She would creep around looking depressed and once she took my phone charger not telling anyone thinking it belonged to Tomas. One day she sort of slid in the room we were all working in not saying anything and I referred to her as a creepy cockroach.

Next day we went shopping for food at a super market for everyone and she insisted on buying their own food separate in which I really lost it and blew up at Tomas knowing having items in a communal refrigerator marked as ours don't touch will not work. I had experienced these sorts of things before once when my brother Bill and his wife lived with us and roped off their section of the refrigerator. Tomas and Sella decided to leave that day got a taxi and rented a hotel for a few days.

We had Tow Boat USA come get us at the entrance to South Cut just before high tide. They towed us through the cut and into Largo Sound at a route I had picked after years of survey. We pulled the Tee Craft behind us.

It was dark when we ran aground and in addition about the same time the Port Engine alarms went off and then the engine stalled. Luckily Sean was there and he had experienced this before in Indonesia and knew what to do he and Johnnie changed the fuel filters and we were running again.

About the same time Mark of Tow Boat USA were pulling us to Port. Sean got in the Tee Craft and helped pull but accidently broke the starboard side engine lower unit gear pin. Luckily the water came up some more, as even though the tide had peaked offshore there was luckily for us a two-hour delay for the tidal peak of Largo Sound, so we started moving again. We got to my dock about 11:00 p.m. Batit and my tenants down stairs had watched the whole thing. The *Princess* tied up well we had a plenty of large piles and plenty of large ropes to tie them to. I was relieved we had made it.

Tomas returned to get his things off the boat and we discussed things amicably as crazy Sella laid on their car horn screaming out the window for him to hurry. I asked Batit my wife to go slap the bitch but she refused.

15.0

Charter to Fox Studios in Miami

Sam Stoia evidently thought he had a boat pull at a ship yard pulling out a catamaran sailing boat owned by a well-known guy in the Keys named Lance whom owns and operates the original steam trawler named *The African Queen* from the movie with the same name. Sam demanded his lift bags back immediately so he could float Lance's boat into his shallow approach. I removed and returned the bags. I advised "you and Lance will never do anything together your personalities will clash." I met and personally did not like Lance. I found him not friendly, arrogant, and dirty like a mechanic. He was jealous of my big boat and called it ugly. Not long after this Joe Dyll lost two finger tips operating one of Lance's capstan winches and one of his boats burnt to the waterline behind his house.

In mid-February, I was contacted by Artie Malechi, who was referred by Sam, to possibly use the *Borneo Princess* for the TV series *Chum* motion camera shoot. Artie soon came with a crew of 6 others twice and confirmed this was the boat they not only wanted but shooting was to take place by 7th March. John explained we could only be sure of getting boat out of Largo Sound by 3rd March as the tides were favorable and we also needed to reinstall the lift bags.

Sam new Artie Malechi wanted to charter the *Princess* and we needed the bags back so Artie and I worked on Sam whom sold these bags to me for USD 2,000 and they were USD 8,000 on the internet so he gave me a great deal something I did not understand as he never did good deals like that for me before. I think he was doing this for Artie. Sean had left so Johnnie and I installed the bags.

Artie was a longtime friend of mine dating back to when we were about fourteen years old. He had lived close to me in Miami Springs in fact only a block away. He was one of the first to get a motor cycle a Honda 90 when he was fourteen years old. Other friends like Robby Knuck, Bill Johnson and Bret Christopher all had small motor cycles before me which made my father feel like I should be allowed to have one although all I got was an old Honda 50 cc. Later I saved and bought an Italian Galero 125 cc. Artie and I would also learn how to go skin diving and spear fish but usually near shore somewhere. Once with Craig Young we walked at a low incoming tide from Virginia Key about a half mile to Fisher Island. The water was getting high and fast as we worried and waded fast like hell with several hundred yards left to go but we made it and cheated death again.

Artie once drove my motorcycle and was laying down going around a long bend near an isolated no longer in use Chrome Avenue. He hit gravel and the motorcycle slid out from under him, but luckily, he had his helmet on, which saved his life. The rest of him was hidden behind a bleeding torn bunch of dirty clothes. We eventually got cars and girl-friends and drifted apart with him pursuing a career with the Sea Aquarium and then as a stunt man started with the James Bond movie Thunder Ball.

On one occasion we had rented a boat at Lowes Marina where present day Catamaran Boat Ship Yard is owned by Sam Stoia. Artie and I had not seen each other since 1970 almost forty-six years previous. On one trip we encountered a huge hammerhead shark that came up to us at Molasis Reef.

I confirmed the *Princess* would be ready as a prop in the bay and could do some sailing but we had just received it from Indonesia and there was a lot of work to be done it looked a mess dirty and with rust spots around.

In addition, much of the electrical was not functioning properly and the Kort Steering nozzles could not be put on in time but since both engines and generators were new and we had come from Borneo to Singapore 1500 miles without a hitch like this both John and Artie perceived the risk to be without one engine, a problem without steering, minimal. I then trusted Artie and engaged five more people working on the boat to get it ready.

We redid some of the electrical including connections that allowed us to use the 220 V single phase shore power system I had installed at my dock years before. We hired Mexican labor and cleaned the boat and removed all rust except under starboard stays as we ran out of time. We spent around USD 20,000 since it arrived in January to get it in shape.

I met and approved Alex as captain and he accepted the boat. I got all the necessary documents including a cruising permit from Customs Key West. I was supposed to have one before we left Fort Lauderdale they advised me, but I played ignorant and so they overlooked it. All I had to do was notify them of any vessel movements.

Some agreements were crudely put in place and two checks totaling USD 45,000 were received by me from Fox. On 2 March to take advantage of high tides and mild winds we got underway about 11:00 a.m.

I had arranged for Tow Boat USA to tow and guide us out of the shallow Largo Sound and through the narrow channel but this time let mark pick the route. The exit out and through the reefs was without a hitch but it was decided to try and just keep the lift bags on to save time. They came out from under boat in the pitching seas and 7-knot speed we maintained. They were recovered on the boat and were ok no holes and not damaged.

We arrived near Fowie Light and were met by Artie in his boat and a push tug was also there. They would help guide us through some shallow and narrow areas. I was impressed at the ability of Artie to have enough insight and perceive risks and provide backup equipment should the worst case happen but did not. We anchored just outside the Sea Aquarium, and in the morning, they brought the boat into a little harbor at the Sea Aquarium.

That day 3rd March and Saturday 4th there were prop men and film crew everywhere getting the boat ready? Production seemed to go very smooth except film production managers were upset there was no one on board at night to act as security for all the expensive camera and recording equipment so we hired a security guy.

One day Artie's crew removed our lift bags off the deck and I came and got them and brought my engineer as they are too heavy (250 lbs. each) to handle alone and took home to spread out inspect. They had two wood splinters that pierced the heavy polyurethane bag. Probably when dragged on the front wood bow. I looked and looked and finally found and ordered a repair kit and then took engineer back to boat.

On the 24th Friday night our worst fears were confirmed as they were moving the boat back into the Sea Aquarium the port engine coupling key sheared and the propeller shaft came backwards out of the engine room but did not fall out the stern of the boat thanks to a design that had a 3.5 ft long stern tube. Artie acted in a flash and considered worst case scenario which would be we lose shaft completely and boat would most likely sink. In an instant he found a round peg and hammered into the stern tube stopping all water but it did damage the seal and leaks more now. The cause of the failure you cannot say operator error as the captain needed to constantly in gage engines forward then reverse to navigate the narrow entrance especially with the winds severe at times. You cannot say bad design as boat had operated for four months without an incident like this and sailed through three seas. My engineer, Johnnie, as I called him then, reported someone had left a large rope in the water and this wrapped around the propeller causing the damage.

I went to visit boat in the morning with Artie and we removed coupling and inspected damage and we believed with some cleaning, a new Key which we had and some new set screws it would be like new so they and we fixed it on the following Monday March 27 at site. I wished to cancel the charter and wait until I went to a ship yard for proper repair but Artie talked me out of it. That weekend March 25 and 26 we could not charter boat out even if it were free of set props so I did not bill for it. Everything ok March 2, 28, 29, 30, 31 and April 1 and 2.

On Friday 31 March I went out with my sailing captain Kim and we inspected all the sails and rigging and found everything all intact. I had met Kim Duncan, a very experienced sailor who was living in a 45-foot sailing yacht on the canal at Marina Del Mar. Kim and his wife Bernadette had a canvas repair and sewing shop where I met him located in the U Haul facility on US1. Kim came out one day and inspected the *Princess* while it was still at Rodriguez Key and said he would like to sail it. He also showed me how to download and use "Sail Flow" a great weather forecast system, which I used religiously.

We were aware the hydraulic winches were not working another electrical control problem so we made sure we could operate by hand, which the normal way for most boats, in the event the winches were not working by Monday, which they still did not, the official sailing day.

Artie had requested a crew but we had no additional people familiar with the boat and in addition additional expenses I had to worry about getting paid for so it was just Kim, Johnnie and myself.

Monday was a steep learning curve it took a little longer to get the sails up but with the help of two set men and Danny it was accomplished. The port engine drive shaft failed again and we fixed it this time adding duct tape as well which saved it from going out the engine room as it failed a short time later.

Artie engaged the push tug at the stern but coming back heading north which we did several times Captain Alex found out he could steer with one engine the way our sailing captain rigged the sails.

At one point the jib encountered a huge wind gust at a worst time when we were coming about (changing directions) it caused the sail to whip and ripped about 1 foot of sail at the base. We decided to furl up a few turns to strengthen it.

Next day Tuesday 4th April was last day and no sailing was required and I was not there and understood they used the push tug as second propulsion.

On the 5th, 6th, and 7th the black and red wax was removed from the saloon roof and deck pretty good but still shows on the sides, stern and on our large boat bumpers. I also noticed hull damage on the port side near the engine room. Our Borneo Stickers had been removed three of including the one on the stern which is a Coast Guard legal requirement.

On the 4th April a company Artie worked with wanted USD 2500 to weld our shaft to the coupling and charge us which I thought was not right. On the Thursday on April 6th, I organized a professional dis similar metal nuclear welder named Joe Hardy who had a tig system and obtained the correct special rods and knew how to control the heat so he did not warp the shaft. We paid him USD 400. On 7th April the electrical part to repair the sailing winches arrived and we installed it. I had met Joe while trying to get my Tee Craft port side propeller repaired who was at the propeller work shop doing odd jobs for the owner.

I picked up Captain Alex and returned to the boat on 10 April to walk through with Artie and see it off to Key Largo. The port engine batteries were low and we could not depart. Artie liked to blame our engineer Johnnie but I don't know maybe he is correct but the engineer says different claims it was an accident.

I had planned to have my Tee Craft starboard engine lower unit repaired, the port side propeller repaired and starboard rub rail repaired and also clean and paint my boat bottom urgently as I needed to receive the *Borneo Princess* in a few days. I had met a Cuban friend of Sam's on site and he agreed to do the fiber glassing of my rub rail area for USD 500 and then I get a detailed proposal saying the boat had to go to his yard and it was a day rate job no mention of the $500 we agreed at Sam's site so I assumed his office management people sent a standard agreement.

I turned this proposal into a works order and emailed to him for signature, which is how the whole world except Key Largo does business. Evidently, I have found out the average ignorant public person here just accept written quotes from contractors never modifying or negotiating terms. The guy got angry not reading what I had written and we ended the conversation with me saying Fuck You. As I was finishing the painting the Cuban comes over to my boat and wants to start a fight. I was not mad I had no reason to bust his skull or put him in jail for striking a senior citizen person namely me so why risk getting hurt he was much younger and fitter so I just talked my way out of it.

I finished the painting and Larry replaced the propeller but there was no time to repair the lower unit so I launched the boat as I had to take everybody off the *Borneo Princess* in a few days.

16.0

Back to Rodriguez Key

On 11 April Kim came with us and we motor sailed back to Key Largo. Kim picked a route in Hawk Channel first time I had navigated it before we were always outside the reefs. We had a good wind 15 knots blowing steadily from the north and so Kim put out the jib sail. With both engines running, we were making 14-knot speed. Everything was great until an engine alarm went off on the Starboard side that was no fuel. We knew we had lots of fuel in the bow tanks and after inspection found the starboard day tank was empty, Luckily Sean showed Johnnie how to transfer fuel.

We had a problem that the engine would not start back up however. Kim went into the engine room and I rang Caterpillar and together we worked out we needed to prime the ejectors which we did. The entire time took about an hour, and unfortunately, we were also entering the narrowest area of Hawks Channel and with only one engine and no rudders I navigated us through. I would put the port engine in reverse which would cause us to turn to port and then once about 20 degrees to port of course I would put back to forward and we would slowly come around on course and a little to starboard after about a mile or so. I repeated the process several times getting us through dangerous waters. I noticed that even with only one engine we were doing 10 knots so knowing with two engines and no sails we would do 8 knots we were moving 6 knots from the wind speed alone. I was learning to be a sailor.

We made it back to Rodriguez Key in the late afternoon and I had arranged for a neighbor friend Jason Poucher to pick Kim and I up and take to Kim's boat. Johnnie stayed offshore as boat sitter.

I had met Jason and his lovely wife Claudia while I was inquiring whether a 20 ft trailer in front of his house was for sale. They had successful businesses in the selling of automotive retail properties in Texas and had just moved here and purchased two houses and two boats. I invited them for dinner one night and we also discussed them to act as an agent to help me sell my CNG business.

Before leaving we tightened all the stern tube shaft seals. Kim took me home in his car. In my tired state I had left my phone on the boat and my car was at the Sea Aquarium. Next morning, I found my computer was not working so could not even contact people on Skype. I had no contact with the outside world and no car. Luckily, I was good friends with Alice and Tuck at the end of our block and they loaned me their car for an hour and I was able to catch a ride with Sam Stoia to the Sea Aquarium whom was going to Miami anyway.

Next day from my house I went out in my Tee Craft with only one engine running, something as a rule I never go out with only one engine, the ten miles to the *Princess* and retrieved my phone and brought Johnnie some food and cigarettes.

17.0

Rodriguez Key Refurbishment

I would leave Johnnie out on the *Borneo Princess* sometimes for a week and then would bring him cigarettes and food and would stay for a few hours adjusting anchor patterns in preparation for storm warnings I would get on my I phone via the App "Sail Flow."

We would also routinely start the engines and generators and check the bilge pumps. Johnnie had been out there long enough and so I was planning on bringing him to land for a week or so.

I was supposed to also take my Tee Craft to Sam's place for an out haul so Larry, my mechanic, could replace my broken starboard lower unit. I had dropped my car off near Sam's across the street at Ballyhoos restaurant and was given a ride back to my house by one of my vacation rental tenants that would frequent the lower apartment on the water.

The problem was what to do with Johnnie as Sam had a falling out with him and did not want him on his property. A large storm was brewing, I had only one engine, and it was low tide making access to Sam's very difficult. I successfully berthed the boat and tied it off and Johnnie headed out of Sam's property quickly. Too late, Sam drives up in the few minutes since we arrived and really "spits the dummy," Australian phrase for someone throwing a silly tantrum fit as a dummy is the name used for a baby's pacifier. I explained the situation but Sam did not care he simply would not listen and demanded I remove the boat from his property saying I did not respect him. I said, well, okay can I wait until after the large storm. He agreed so I took Johnnie to my house and came back after the storm to retrieve the boat by myself. The wind still blowing from the SW as I untied the lines the wind took the boat as I made a swan dive on to the bow deck cracking two ribs. I was able to get it started and avoided colliding with any other boats or large concrete walls. My ribs were sore for months. I considered suing Sam for the injury. Johnnie helped me close the Ocala factory and bring all the equipment in two trucks to Key Largo where we stored in a U Haul Storage facility. My overheads with OES CNG SMARTGAS now went from USD 10,000 per month down to USD 250 per month.

Sean had flown out to attend a Monroe County hearing on April 22 as technically he was owner of 603 Island Drive. The hearing was about me, and twenty other people there breaking county ordinances, using my apartment as a vacation rental. It became obvious to me they had no real evidence and really did not care it was just about getting more revenue so I asked what am I allowed to do and they said rent it out for a minimum of a twenty-eight-day lease and so I pleaded guilty accepted the USD 5,000 fine after all we were earning USD 35,000 per year from it and just moved on. I delisted the unit from Airbnb for a while until the heat died down and also only did twenty-eight-day leases.

Sean and I and Johnnie removed the Port side propeller and drive shaft without too much difficulty, standing on the sea bed breathing air through the Hooka Rig we had. We then delivered it to Milan's Machine shop in Doral, Miami.

A machine shop manned by Cubans but run by a European named John Paris that did not speak Spanish. They had previously repaired by welding my two steel 36-inch diameter Kort nozzle rudders that were ripped off in the voyage from Balikpapan to Singapore.

We meaning Sean, myself, Johnnie, Milan's Machine and other experts we met with generally agreed we had to re machine the thread on the drive shaft so the large screwed lock could be tightened down and therefore hold the shaft to the coupling when reversing. In reverse the shaft rotates clock wise and wants to also turn into the lock nut so to everyone this would solve the problem. No one thought it important that when you move forward the counter clockwise rotated shaft wants to unwind out of the nut as both the shaft and nut are turning together. Strange unpredicted things can happen as we would find out later and I wished I would have drilled a hole through the nut and shaft and locked it with a pin which was briefly considered and not followed through with.

We also replaced the port side packing gland and thick rubber ring Sean made from sheet rubber we had bought. No need to touch the starboard side, or so we thought, as everything was fine, no problems ever encountered; we never knew we would be fleeing a hurricane and in addition we were in a hurry Sean was leaving soon and there was simply no one else or place where this could have been repaired. No shipyard could take the boat our 100-foot-high mast could not fit under any power lines or bridges except one in Fort Lauderdale named Directors and they claimed our boat was too small they only help big customers. There was a company in the Bahamas, but I did not want to make the trip without the nozzles and repairs, and once we made them, we did not need a shipyard.

Sean and I also fitted the two Kort nozzles however it took about two days per nozzle. What we had built was from original design drawings but there had been evidently some field fitting when the nozzles were originally fit and changes never made it to the drawing board. They did not easily fit up and they were heavy about 250 lbs. each and had to be lowered using the davits and rigging lines tied and push and shove while Sean and I stood on the hard seabed. When they did not fit, it was recover and cut and grind something and try again. After four days, a day before Sean was to leave, we finished. In the process of being underwater for most of four days and in rough seas at times I developed pneumonia and was coughing a lot but did not know how serious it was becoming.

After Sean left at the end of May, I was in the water coughing, trying to get the rudders in line. They were off by about twenty degrees each aimed out about ten degrees, which was not too bad, and we went with this thinking we will fix when it is behind my house. We only needed to install the shaft and Kort nozzles behind Rodriquez Key where we could stand in clear water.

One night, I coughed so hard as I sat up on the side of the bed that I passed out, falling headfirst to the floor and landing on my right eyelid in a way that also bent my back wrong.

I remember waking up in a pool of blood on the floor not knowing or remembering anything. Batit was very scared of all the blood and I could not talk for a while. She got me up and wiped the blood off trying to bandage a huge gash. I was more concerned with my back as I could not move much and thought, *shit, I have broken it.* I stayed in bed rolling over or getting up to pee occasionally with great pain having to work out the least painful way. When I was able to get up, I could not walk alone; Batit had to walk me. Next day I was a little better and went to the Clinic to see Doctor Ray whom prescribed something to make me stop coughing and an antibiotic and sewed up my eye. Best of all it, cost me nothing as I was now for the first time on Medicare. Yahoo! I have made it to old age and cheap doctor services.

After a few days I returned to bring Johnnie food and cigarettes. Johnnie was actually doing good he would catch fish including a large shark and he even had two girlfriends he met going by on their jet skis whom he would wave down and they would stop in and take care of his sexual needs. Johnnie would also get boat visitors including my friends Niki and Carlos.

A few times and once when I was there, we invited and were boarded by the Coast Guard, Homeland Security, and the FWC whom were all curious, friendly and loved the boat. We had made friends in all the right places.

We called Mark at Tow Boat USA and met him at the entrance to the South Cut and towed us to my house this time with no grounding as I followed his route advise same as we had used to get out the last time.

18.0

Cay Sal Bank Charter

I had made contact with a Steve Moore of Mobile Diving Key West whom which I saw an ad appear on my lap top while on the internet. Steve visited the boat and immediately fell in love with it and wanted to charter it to Cay Sal Banks. Steve was a young sixtysomething ex-medic in the Coast Guard originally from Louisiana. The Cay Sal Banks is a shallow water live coral reef system and also has few uninhabited group of tiny islands north of Cuba but belonging to the Bahamas. The banks are known for their abundance of coral, fish, sharks and large spiny lobster. The problem is they are ninety miles from the Keys and you have to cross the Gulf Stream the world's largest and deepest river. The weather can also blow up and there is hardly anywhere to hide. To go there you need a big boat with accommodation and so the *Borneo Princess* was perfect. We would tow his 25-foot boat behind as we needed a mobile dive platform.

Steve agreed to charter the boat for the cost of fuel plus some extra fuel for me, entry permits into the Bahamas and provision of food. We formulated plans in mid-July with departure set initially for 1st of September. We were to stay for ten days and first clear into

Bimini, Bahamas. Steve agreed to supply his dive compressor fitted with a new 3 phase 50 hz motor to match the vessels generator output.

I had to get the vessel ready. We already believed we had all the drive shaft and steering rudder problems solved. We needed to get the engine room bilge pumps all safely operational as what was done by the previous Cuban electrician was not good as he used 25-amp breakers for a 9-amp pump. We also had to get the air conditioning system functioning which had never worked and the sailing winch hydraulics also had failed. Three track plotters were made functional with only the addition of sim cards so we could see where we were and water depths with the GPS systems. Two were Garman the third a Simrad system. The Simrad automatic pilot system received signals from a 12-volt signal bar as well as the other display devices where all instrumentation systems including position, heading speed, wind speed, radar images could be obtained. A new 12 Volt VHF radio was also required and there was no reliable 12-volt system on board so we had to get dedicated 12-V batteries. The previous engineers had tried to get 12 V off the 2 x 24 V house battery system.

Johnnie Rodriguez was not suitable to cope with the challenge of any of this electrical and instrumentation work. In addition, his immigration status was at best dubious a refugee from Cuba that had served time for a felony but could not be deported as the US and Cuba had no diplomatic relations so he could easily be arrested on reentry into the US, so I decided to phase him out with a new crew. I met at Lonnie's, a good friend of my wife Batit, Justin McKnight whom had just moved down to the Keys from North Carolina and was living on a boat behind her house. Justin was an electrician / dive instructor and was keen to go to work for me. Later his friend Michael Maffee originally from New York whom also lived in North Carolina was a professionally trained electrical technician and had private boating experience.

They were to be my crew and I issued them contracts they got USD 125 per day for a minimum of five day week and offshore this was increased 50%. At first Justin was to live in my downstairs apartment for USD 350 per week but his wife did not want to stay in the Keys so left. Michael then took up the offer and his lovely girlfriend Anna came down and brought his two-year-old son Oryan whom he recently gained custody of.

We had five weeks to get this boat ready and there were other problems as well like the port side grey water pump which was 220-Volt broke down and you could not find 220-Volt pumps and the Port side engine alternator would not put out enough voltage to charge its own battery system. I wanted to clean all the wax left on the boat by Fox Studios so bought a large high pressure water blaster. We wanted to pretty up the boat so I wanted to sand and oil all the teak deck, transom and hand rails. The wheel house floor needed sanding and marine varnishing. I wanted better furniture than what we had in Miami and found it at the Salvation Army.

The Salvation Army Key Largo regularly got donations from Ocean Reef a gated community occupied by billionaire's children the heirs of Campbell Soup, Heinz foods and many others. We had to replace the 220-V refrigerators with a 110-V one so lucky we found on board a 3000-watt transformer which I was told belonged to Tomas. All of the 220-V fans were also replaced with 110 V fans and luckily found and bought 12 x 110 / 220-V small transformers for USD 15 each. We also had to design and install a 220-V 3 phase power supply for the new dive compressor. A detailed preparation schedule is given in section 11.1 below.

Johnnie was being phased out and so he luckily found and I bought him a $2500 RV so he had transportation and a home with air conditioning. All he had to do was earn it so I had to find him twenty-five days' work in which to do that. As I expected he had huge conflict with the Justin and got in a huge argument with him. I sent Johnnie away to help Monty paint my Redlands house where he could live in his RV and repay me with work.

Steve organized 1000 gallons of diesel to be delivered He also had his brother Bob bring from Texas 6 x 55-gallon empty drums that he filled here with gasoline for his boat.

In the meantime, we were very busy Carlos a senior electrician had designed, ordered and installed the compressor supply and also wired up the new compressor motor. Justin worked well with him.

We had a marine hydraulic company named Nance Underwood out of Fort Lauderdale come out to assess our winch problems and much to the company's lack of credibility and overcharging tactics I had to agree to pay $600 for an investigation including travel time of three hours for a young technician named Jonathan to come have a look. I wanted to discuss the problems with Jonathan but the manager an asshole Texan named Houston Murphy refused to allow me to talk to him then gives me this outrageous bull shit quote of $1800 for two guys whom I assumed were seniors to come fix the problem.

Jonathan comes out with a young trainee. They fixed the problem in less than an hour and a half with no parts required and I should have sent them on their way but thought would have them look at our hydraulic steering winches. I was away so could not talk to Jonathan directly but did manage to get his cell phone number. I rang Lance Underwood to find out what they had done but got the asshole Houston Murphy with no help saying just pay the outrageous bill of USD 1800. When I added their time including travel and a charge out rate of $45 per hour each representative of the age and experience, I got $700. I then rang Jonathan to ask what he had done and left us with and he advised he would be fired if he talked to me. This lack of understanding of what we were left with our hydraulic steering system would contribute to the sinking of the *Borneo Princess*.

One day around 25 August, Michael came running to me threatening to quit claiming Justin had abused him and assaulted him. I listened to both guys accounts and decided they could not work together anymore and I sided with Michael as he had proved a much harder worker and more valuable than Justin and so let Justin go. I paid Justin for his time, wrote a good letter of reference and also took him and helped him rent a car to return to North Carolina.

Steve his brother Bob, son Justin and a young oil worker dive friend Jason all came and stayed on the boat starting on 1st September helping get things ready. They had to carry the 400-pound compressor and the four large oxygen bottles as Steve was proposing to do deep diving to 200 ft and so by using an oxygen content of forty percent which is twice of that as normal air and would provide an equivalent nitrogen partial pressure as someone on air at 150 feet. In short you were less likely to get nitrogen narcosis or the bends. The partial pressure of the oxygen would be however 2.42 bars gage pressure, which could cause oxygen poisoning, so I assume he would have limited his dives to less say 180-feet. Jason, a very young and fit young man, proved a huge asset as he almost single handed climbed our 100 ft high mast tower and installed the 30-pound mast head housing all the necessary navigation lights. Bob also brought steaks from Texas.

We had meetings with Steve Levine and his friend, Tony Marchetti whom was interested in coming. I met Tony who was a retired swim instructor from up North who went out diving with us on Steve's 30-foot boat a couple times. Tony was an exceptional free diver could easily hit one hundred feet and stay down for minutes. He was new to the keys when I met him and new to spear fishing but learned fast. Steve wanted to bring his Jet Ski and so we tried to do a trial loading using the lifting davits and sail winches for pulling in board. It did not go well and we nearly broke the Jet Ski lifting ring. We could have done it with two lift davits and two pulling winches but that meant four men and Steve Moore did not wish to provide his men for that any longer. I text Sam Stoia at Catamaran boats to see if he would help pull out and store Tony's boat during the hurricane which he did.

18.1 Story of Steve Levine the Jet SKI King

One day about three years previously a guy came speeding by my house and dock breaking all the County speed laws about protecting docks and also sea grass a source of habitat for most fish and lobster. I ran out yelling at him and he came back to see what was wrong with me. I explained he understood never did it again and we became good friends. I did not know until later he was a paraplegic. Steve a real-estate salesman was diving in a cave using a hooka rig or hose supplied air in 2001 and his weight belt came off sending him to the surface and too fast and he got what is called a central nervous system hit when an air bubble forms and cuts off the flow of blood to that part of the body. He was not treated properly by the charter boat or especially the doctor and was not put into a recompression chamber for 24 hours. He successfully sued and got some compensation. He later moved to the Philippines and opened a bar. Once there was an attempted robbery at gun point and Steve foiled the whole robbery by grabbing one guy and taking his gun and sticking it in his mouth causing all the thieves to run. Steve moved to the Keys in 2010 a few years before I moved here. He would take me out and taught me how to catch lobster and we would also look for lobster holes as he would tow me behind his jet ski. I would spot a school of fish swimming and follow them to their hole it was that easy.

Steve even without legs after spotting a lobster would dive down to ten feet water depth using only his arms and hands and would hold on with one hand and feel his way in the hole until he could get his hand safely around one. He would often get more than me. He taught me how to use a tickle stick to get lobster to leave hole and then net it which worked good under many circumstance but if the lobster had a huge hole to back up into this would sometimes not work. I learned to also stick my hands in but once bitten by a Moray Eel gave me second thoughts.

18.2 Cay Sal Preparation Schedule

Week 1 July 24 to July 29
Tee Craft bilge pumps
Borneo P. bilge pumps
Back deck sanding / oiling
Buy back deck furniture
Install new refrigerator
Refurbish Air Conditioning System
Anchor rack and Anchor repaired
House Batteries and solar controller tested
US Customs Key West visit for phone in authorization
Check all pumps and electrical facilities

Week 2 July 31 to Aug 5
Arrival of Michael and Steve
Review Instrumentation and Auto Pilot systems

Johnnie R to show Michael and Steve how to start engines and generators and fuel transfer JR Connect new 220 v 3 phase motor to Steve Compressor

Pick up and load ne furniture remove black table and old couch JR. Michael Passport and pick up anchor in Miami

Install anchor and anchor rack JR
Sort out wires up mast, label and test

Pay vessel registration
Fill and relocate lift bags away from AC cooling water discharge JR / JL Buy and install printer

Prepare an instrumentation and electrical booklet JL Order Port Grey Water capacitor JM Remove Wax from sides of boat JR

Investigate Port Fuel pump JR
Find Sea Anchor JR

Week 3 Aug 7 to Aug 12
Clean BP Stainless Steel hand rails JR
Coat hand rail wood with teak oil JR
Install Mast Top
Install Radar

Clean hydraulic spill, tighten fittings get hydraulic tech. to test and bleed system Get AC man out to clean / test water cooling system JL

Adjust stern rudders JR / JL
Advance automatic steering system
Get zodiac boat and motor from Redlands
Purchase 6 220v fans

Week 4 Aug 14 to Aug 19
Automatic pilot completion
Install 3 phase 220 v for compressor Carlos / JM

Clean and oil all stay wires
Water maker commissioning
Build venturi as back up to empty black or grey tanks JL
Strap bow wood panels down JL / JR

Scrub boat clean all rust JR/JM
Install new decals JR/ JM
Get new Indo / US wall socket adaptors JL

Week 5 Aug 21 to Aug 26
Steve to arrange 1000 gallons diesel fuel SM

Ck engine oils, review change schedules get oil if required JM Add 2 tonne water 530 gallons spit each side JR

Transfer 1 tonne from port green to port blue JR / JM
Notify Tow Boat USA JL
Provision of food JL / SM

Bedding JL

Week 6 Aug 28 to Sept 2
Load Steve Gear JR
Satellite Phone SM
Test Air Compressor SM
Tow out of South Cut
Go

19.0

Hurricane Irma is Coming

By Monday Irma was fast approaching heading still due west rolling up The Dominican Republic not heading North as expected and hoped and so the trip was cancelled. Jason helped us install ten 2 x 8-inch treated beams to stiffen up the piles. Simple calculations I did for wind only showed the piles just ok for 100 mph winds and that did not allow for wave loading which would prove significant not to mention the 160 mph winds which we did get which would give a load of 2.5 times larger.

On Tuesday 5th September I contacted Carlos Rippes, through Johnnie as he was helping him, to see if they wanted to join us and bring his 35-foot boat that we could tow behind us and go to south of Cuba to seek shelter from the storm.

Carlos was a Cuban lawyer I met though Nikki. Nikki named his 25-foot boat *Carlito* after Carlos. They jointly owned a property on a bay side canal with a small modular home built on it. Carlos had bought out Niki's half but before that bought a 35-foot single engine cabin cruiser which he kept on the canal but wanted to move it to safety from Irma. Carlos was a very intelligent quiet and considerate person whom had almost a feminine personality and I consider him a good friend.

The hurricane was still five days away and we could make Cuba in twenty-eight hours. They agreed so we all made the necessary preparations in a hurry. Johnnie came luckily as he proved a good asset and worked well with Michael and rang Ramon a 350 lb. Tahitian, whom I knew a little, whom, also, came. Ramon when he was young in high school was insulted by a French school teacher that said "all Tahitian people were stupid" and so he stabbed her with a pencil in the face.

He was given a choice go to jail or join the French Army, which is what he did. He lived in Paris and other parts of France and grew up. Back in Tahiti later, he met a young American girl and they married and she brought him to the USA. Ramon was one of the most pleasant people I have ever met never quick to anger and always courteous to everyone. He was however at one with nature preferring to shower with the hose and sleep outside. He had a strong addiction for beer and cigarettes as many people in the Keys I would work with.

On Wednesday we found the port engine would not start and Michael was unsure of the problem so I rang Johnnie who had been on the boat for 6 months and hoped he could help. He came and he correctly deduced one battery was bad so we replaced under warranty for free at Napco. Batit was to stay with her friend Lonnie, so we put all the hurricane shutters up and carried outdoor furniture into the bottom apartment. Some furniture would not fit so Michael and Batit tied it all together. Michael's mother came and picked up Anna and his son Oryan and took them to Orlando. Michael had replaced and rewired the bilge pumps for the 28-foot Tee Craft that Justin had screwed up. On Thursday the day of departure we found the electric stove was not working so I quickly borrowed a gas operated one from Mark at Tow Boat USA. We took the Tee Craft down North Cut into the mangroves and tied it to 4 large mangrove roots.

20.0

The Last Voyage

Mark from tow boat came at 11:00 a.m. at high tide and towed us out the South Cut where we anchored and came back with Carlos's boat just to tidy up everything at our homes for any last-minute items. We left Ramon on the boat. We came back and Michael and I and Ramon with the help of Johnnie at the sail winch deflated the two 7000 lb. lift bags and recovered them on the back deck

We were underway by about 4:00 p.m. From the start I thought something was wrong we could not steer correctly then somehow it improved and we made 8 knots in a straight line for Key West in Hawks Channel with Carlos 35-foot boat in tow behind us. From about 7:00 p.m. to 10:00 p.m. I got some sleep and left Ramon at the wheel that did a good job following the route line I had drawn. Ramon had evidently consumed most of the sodas, fruit punch and much of the Scotch alcohol that had been brought on board.

At around 11:00 p.m. we could no longer steer so I went down in the water on the Port side and found the steering Kort nozzle was no longer there. I dove on the Starboard side and found it loose. I decided to try and remove it with intoxicated Ramon holding an underwater light from above which did little good and me trying to hold two small wrenches in the pounding seas. I gave up and we anchored for the night at about 2:00 a.m.

Friday morning. At 4:00 a.m. I woke up and got everybody else up in the night it came to me to try and tie the Starboard Kort nozzle in place with big ropes running to each side of the ship. Michael was first one in the water and he looped the line through one side of the Kort nozzle then through itself and it was pulled very tight and secured to the Starboard deck cleat by Johnnie. I went down put another on the Port side of the Starboard Kort nozzle and we pulled hard with the stern anchor winch. It was done we could now steer by adjusting each engine speed and direction.

At some point there was a burning smell coming from the starboard engine room which turned out to be the gland packing. Michael loosened it some more and the problem seemed to abate. At some point he advised the bilge pumps were not working on that side.

We were underway for a few hours, all celebrating our success when Michael noticed Carlos's boat was no longer behind us. Everyone got up and we pulled in the tow line and went North to find it in the windy rough night. Luckily, he had a light on it and we finally found it drifting and rolling and pitching. We made several attempts to get alongside but the pitching and strong winds would just bow it past and we could not get a line on it. I had to constantly reverse my engines to control my boat and keep from smashing into his boat.

The last time I was prepared to continue with this operation and lucky for all of us Carlos the owner made a flying leap of faith from our deck down 3 ft to his deck and was able to get on board and start his boat. He came up behind and we gave him another tow line.

I noticed our boat had little power and steering was an issue again when Michael yelled "We are sinking" and advised we had four feet of water in the port side engine room. I knew we would not sink and also believed we must have lost the entire port drive shaft out the back something that had never happened before. I got a mask and jumped in and my belief confirmed. I noticed there was no more water entering as there was no suction which was probably a good thing as I could get my hand sucked through the three-inch opening.

At first, I requested some plastic bags and Ramon brought one then I said no need a bunch. Then Johnnie to his quick wit found the one we had used for Fox Studios in Miami and brought it. I did my best to wedge my buoyant body under the hull hold my breath and drive the plug in with a hammer though I was not happy it was in tight enough.

The engine kept running during all of this and the water level was just under the alternator underside. Michael quickly got the gasoline bilge pump in action and within 60 minutes pumped the engine room dry. To his credit he also checked behind the engine room bulkhead in the port stern cabin under the bead and found more water and pumped it dry.

We tried to get underway but could not steer with one engine and no rudders. We pulled out the 100-foot-high jib sail hoping to head west but we were heading directly for the seven-mile bridge and could not turn away. I called the Coastguard for help but they advised they were shut down for the Hurricane. I tried Tow Boat USA and they also had no one in the area and were too busy all fleeing because of the hurricane.

Finally, we opted to abandon ship and make for Key West in Carlos's 35-foot boat not knowing where the hurricane was or where it was going but hoping we would stay afloat and return in a few days. Johnnie and I dropped the two 160 lb. bow anchors that already held fast in 50-knot winds. We also threw out the 160 lb. plow anchor that that held fast in 40 knot winds. So, in theory, we were ok for 90-knot winds not knowing we would get closer to 180-knot winds with a force of four times of that as a 90-knot wind.

We all closed the wheel house windows very tight and checked the saloon windows too. The windows could take virtually anything the sea could throw at it. I grabbed my clothes Batit had packed and tied them together and for some reason my dive and spear fish gear and the barbecue

pit we borrowed from Mark at Tow Boat USA. Michael grabbed his rif-
fle but forgot his tools for some stupid reason. Johnnie and Ramon
grabbed some food and drink and especially the 10 lb. ham a rarity he
had seldom seen. We got underway about 9:00 a.m., headed for Key
West, but were not sure where to go once we got there. I went into the
main cabin confused and tired and tried to sleep but it was too hot.

21.0

Trying to Get Home

We arrived in Key West about noon at the Boca Chica Marina around owned by the US Navy and presumably only for ex-Navy staff officers. There were hundreds of large pleasure craft all tied up with multiple lines for the storm. The place however was abandoned not a person in sight. We found a berth and put out four lines. We then walked along the docks looking at dozens of large boats then out the marina never seeing a single person. We walked a mile down a long road that ended with a locked gate with barbed wire. You could hear the road traffic along US1 which is where we wanted to go. The road also turned a hard right before the gate and looked to continue for miles.

We saw an occasional vehicle far down the road. It was very hot, probably 95 degrees F, and we were in direct sun with no wind carrying various forms of personal effects. It was hard to imagine a hurricane was supposed to be coming. Johnnie and Michael, both fit, were able to climb the fence and negotiate the barbed wire. I tried but failed and had already thrown my clothes pile over. Ramon and Carlos did not even try. Johnnie and Michael, we would learn later hitchhiked to a bus stop and caught a bus north to Key Largo and then Michael to Orlando to join his family.

I tried to break the gate lock with a large boulder several times and then tried to dig under the fence with no luck. Carlos, Ramon and I then continued down the seeming endless road almost passing out from heat exhaustion. Then I noticed far to the right about a hundred yards some men working at what appeared to be a warehouse loading dock. I yelled and waved my arms and kept walking. They came out to meet us a little apprehensive at first with machine guns as to why we were on their property there had not been a storm yet so we could not be refuges in their minds. We told them our story but they were not convinced especially since I had an Australian accent, Carlos clearly Cuban and what to make of 350 lb. Ramon with no identification. They took us in one car to where I said my bag was thrown over the fence perhaps to check our story. A military car followed our car with armed guards. We were taken to command headquarters given lots of cold water and interrogated and gave them our identification with Ramon having none which posed no problem for some reason. They asked me why I spoke with a foreign accent and I told them I had lived in Australia for thirty-five years. They agreed to take us to the bus stop. When I reminded them to please first take me to get my clothes a plain clothes CO whom had just arrived said I was very demanding. I said yes, I was General John B. Lincoln he laughed. Satisfied we were no threat they eventually took us to a bus stop where they advised there were free evacuation buses from the Keys.

We sat at the bus stop for what seemed hours with no bus in sight. One of us occasionally stood along the road and hitch hiked but this seemed to irritate some drivers whom blew their horns and gave rude stares as if we were homeless idiots not worth their company. One Cuban lady showed up and said there were no more busses running and then she said she was waiting for her brother to come and for USD 100 we could catch a ride to Key Largo. She ended up getting on the bus so I assumed she was crazy like me.

A bus did come and it was free but took us on an hour tour of Key West before going to the bus terminal where we transferred to another free bus and at about 7:00 p.m. landed in Key Largo and I had the driver drop us off across the street from the U Haul dealer where my car was parked. I drove to Carlos's house and Ramon made some spaghetti sauce and pasta and we ate after first getting a little drunk on Scotch whiskey. Carlos and I went to Lonnie's house on the Bay side not far where my wife Batit was going to wait out the hurricane. They gave us some rice dish with clam sauce that we took with us and left soon after.

Ramon and I headed for my house to ride out the storm. It was dark and we had a hard time getting in the upstairs front door in which was closed with the shutters and we had no tools. Once in I hooked up my computer to the internet and followed Irma it was still 450 miles east away along the north east coast of Cuba rolling along the coast and it appeared to us it would continue west and hopefully miss Florida altogether. Saturday morning came the winds had increase to about 35 knots out of the east and the hurricane still hugging the coast of Cuba but now closer about 300 miles east.

We dropped my 2010 Ford Explorer off at the U Haul yard as it was at a very high location about 15 feet above sea level and unlikely to flood. We walked back in 40-knot winds getting a lot of strange looks from occasional passersby. We took a few photos, joking around and in one scene I stripped naked along my dock and posted it on Face Book saying my clothes had blown off which was a favorite expression of Miami news reporters although the winds there never got very large.

We prepared the best we could. I filled one bath tub up with water for toilet flushing in the event we lost water pressure. We cooked all the food we could easily find like hot dogs, a dozen eggs, two lbs. of bacon and five pounds of shark meat Ramon found in the freezer from which he made a stew.

We still had power and internet and I could track the storm on Google Earth. I noticed the storm after closely following the north coast of Cuba suddenly kicked north near Santa Clara a large bay. We lost internet and cable TV not long after that and so no longer knew where then storm was or heading. If it continued north, it would have missed Florida and hit the Bahamas but evidently went North West hitting south of Marathon exactly where we had abandoned the *Borneo Princess* before turning north east and heading directly for Key Largo.

22.0

Hurricane Irma Hits the Keys

At midnight the water started to rise and the winds blew harder. I woke up with water droplets hitting me to find the fan was spinning and throwing water around the room. Evidently there was a rain leak and I suspected the worst in that we had lost roof tiles. There was also water dripping in through the roof in the living room. I spent many hours moving electrical devices and placing buckets and towels around. Just when we thought we had the problem licked we hear a gush of water enter the house and assumed the worst that the water had risen 10 feet and was coming in through the stove air vent.

Thank God that was not to be the case, but for some strange reason at the time the washing machine filled up with water and the door blew open and spilt about five gallons.

That was weird we shut the water tap off so we thought and had to scrape the hall and my bedroom with a dust pan and bucket and recover five gallons. We had a partial power outage I believe as the downstairs stove shorted out, the only downstairs electrical outlet we could not find a breaker for which also located low near the floor, from the rising salt water. That night I felt like shit to say the least. Ramon's shark stew tasted bad and made me sick.

The next morning on a Sunday the wind was even worse switching from the South East more now from the South and estimated to be 150 mph, was howling and water kept coming in from the roof and from the fucking washing machine even though the water was supposedly turned off. We spent the morning dust panning water, moping and ringing out a dozen towels. We had no idea that we were very close to the eye of the storm about forty miles as it passed north west of us heading north east for Naples about sixty miles away.

We went outside upstairs on the west and front veranda to the lee side of the wind. We watched in amazement at the storm surge in the backyard about ten feet higher than normal with large breaking waves and photographed it. We also photographed at first our mail box and the neighbors washed away in the white raging rapids about four feet deep which used to be our front yard and street. Then came the 5-foot-long wooden buffet table followed by a large wooden table with six chairs tied to it as it went out our gate across where the street used to be and through the neighbor's yard across the street. A large sign posted "Slow Down Dead End 5 MPH" somehow got erected on our fence right side up and later never to be seen again. I texted Kathy Klock, our neighbor, these photos plus one of Dick Malgaldie's house, our neighbor to the west, the front of which showed no damage.

23.0

The Long Cleanup

By late afternoon the water had receded and the wind died to a comfortable 40 mph. We got a ride with a man and his daughter in a pickup truck to get my car. My car and truck were okay, still there and had not been flooded. We still had power but had lost the internet and cable TV. We went to Lonnie's house, and as I walked up the external stairs and knocked on her shutters, she came out screaming, "What do you want? We don't want you here!" I said, "I have come for my wife." Batit stayed there and we returned home with the car but had only five miles of fuel left, according to the tank gage read out. We still had some water pressure and so could still take a weak but warm shower.

Ramon and I went for a walk down the street and noticed Kathy's 20-foot boat was in her house. Ramon also found twelve full beer bottles on the side of the street, which was a nice find as we had no alcohol. That afternoon the power failed and we lost our ATT phone signal as well.

The next day, Monday, the storm had passed for good, heading north. After a breakfast of cold coffee and peanut butter sandwiches, we took a walk to the east along the water's edge in everybody's back-yard to find every boat's supposedly strong aluminum modern boat lift facility had broken and their boat had either washed up in their back-yard or had sunk. Kathy's boat had smashed into her house.

Dick, my unfriendly neighbor for seven years to the west, came over and shook my hand as he had stayed at a shelter and was happy I sent a photo of his place undisturbed to Kathy, who had forwarded it to him. He even helped plug some of our PVC water leaks along our washed away dock. Even more amazing Dick let us borrow his car in search of a gas station open but found none He loaned it a second time for me to go visit Batit, but there was too much water on Lonnie's Street to drive down, so we went to Ramon's van parked behind the Bait House and picked up his cat Nate.

A guy named Michael came by in a Jet Ski and I waved him down. He gave me a ride out South Creek into a shallow channel in the man-groves to see if my 28-foot Tee Craft was okay. It was fine, somehow survived.

Ramon had somehow managed to find all the pieces to my Weber Bar Be Q and so we found old lobster traps and broke them up and made a fire to cook hamburgers and fish that had thawed out in the freezer. We still had a little cold Coke left to wash them down with.

Kathy's neighbor two doors down, a somewhat skeptic antisocial lawyer, ended up with his boat in his new swimming pool. His imme-diate neighbor to the east had a $300,000 Terra that sank. I asked him if he needed help as I had a guy Ramon working for me for room and board and was inexpensive and could use some cash.

His reply: "I don't want any petafilers as I have two daughters." I replied, "Ramon is not a petafiler." His reply, "I told you I don't want any petafilers." At this point I walked away thinking what an asshole.

Ramon and I opened most of the metal sliding shutters upstairs, which protected us throughout the storm.

We noticed there were a total of five open ended PVC water lines at every one's broken dock, which were also broken, and so we turned their water off, closed nearby valves, or plugged them. This increased our water pressure so we could now take a stronger shower, albeit a cold one. In the 90-degree heat, with no wind after the storm the cold felt good. I would wake up four times at night and take a cold shower to cool off.

Ramon, a true nature's child, preferred to sleep outside on the back porch and showered outside with the hose. He would train Nate to go down the stairs and do his business outside. Nate was more like a dog than a cat as it came when you called it and loved to be petted. It also would eat anything including salad and bones.

On Tuesday morning after cold instant coffee and peanut butter sandwich for breakfast, Ramon had found a job at $13 per hour working for Dick, so my work was put on hold. Another guy a friend of Steve Levine's came around on a jet ski and I hitched a ride to get my 28-foot Tee Craft. Tee Craft started right away, and so I untied the four ropes tied to large mangrove roots. Two of them to the south had broken from the trees and were dangling underwater. The mangrove roots to the north themselves acted as a cushion and held the boat.

Kathy drove down from West Palm Beach and came over and brought flashlights, Sterno canned heat, batteries, and a hot chicken, which we consumed for lunch. Another previously unfriendly Russian neighbor named Igor, probably as I have yelled at his children and him for speeding past my dock in their Jet Skis over the years, came by and brought some breakfast cereal, oatmeal, pasta, and a box of candy bars, which we readily accepted as we had no way to get food as there was no gasoline in my car even though all grocery stores had not yet opened there were some Thom Thumb stores open.

Later Kathy's nephew's friend Jason, a web site designer, took me to Taveneir and we found gasoline however there was a long line of cars. I filled up two five-gallon cans and gave one to Dick whom had paid me $20, when we returned. I tipped Jason eight dollars. I put five gallons in my car, which gave me an additional ninety miles to the five I had, and then went to Tom Thumb and got some ice, beer, juice, and milk.

I went looking for my wife Batit and drove to Lonnie's as the water had subsided. She was not home. When I returned home, she was waiting outside with suitcase in hand. We had a happy tearful reunion and made love that night outside on the cool front porch.

She was very unhappy with Ramon having access into our house as without air conditioning, we had to leave all the sliding doors open and she worried even more as he drank a lot of beer each night. I never worried. I trusted Ramon.

Wednesday morning, we had hot coffee thanks to the Sterno Kathy brought along with cold peanut butter and jelly sandwich for me and hot oatmeal thanks to Igor. We then started the painful, as you don't like seeing your damaged things, cleanup, and throw away of all the damaged downstairs furniture. At first Batit tried to take over and was clear to Ramon and I she was only getting in the way. We took outside everything and threw to one side three damaged couches, two bed tables, one lazy boy chair, two chest of drawers, a roll out single bed and the box springs and mattress of our queen bed. We could not find the power breaker for the stove, so couldn't turn it on and test it. The refrigerator worked well. Ramon washed the unit out with fresh water. I soaked all the solid wood or rattan furniture that we decided survived except for the cushions. The cushions from one L shaped large couch that was usually outside had been put high up in the apartment and were not wet. Luckily as these would replace most of the damaged rattan cushions as if it had been planned that way.

Ramon and I found some more fuel in Tavenier and briefly attempted to drive south to the Seven-mile bridge to look for the *Borneo Princess* but were turned back by state police as bodies were still being recovered and it was closed to everyone.

I took the 30 HP new outboard engine, that was intended for our dingy being shipped from Singapore, out of the utility room and had been submerged in fresh water using our large blue recycle bin. I made a noodle dish using the pasta the neighbor Igor brought over with some red tomato sauce we had in the freezer. We went to Winn Dixie which was now open and bought some ice, pork ribs, canned bake beans, cheap beer and salad material. We made a fire in the Weber Bar B Q with table extension boards and Ramon cooked the ribs and Batit made rice. We had a good meal our best in five days.

At 9:00 p.m. our mobile phone service was restored but we had no power to charge phones so we asked our neighbor Dick next door to charge them as he had a generator running. Sean my son from Australia rang that night to see how we were doing. My daughter Melissa also texted me. Took a cold tub bath that afternoon and for some strange reason Ramon came in to see me for which I told him he was off limits

Thursday, we made hot instant coffee and two chocolate bars for Ramon and peanut butter sandwich for me again. I found out peanut butter sandwich is what you were fed in jail and prison and Ramon had done some time for robbery of a jewelry store and never wanted to eat them again. Ramon and I cleaned out and washed out the utility room another depressing job for me as I did not want to see all the damaged items. Michael the same guy that took me to see my 28-foot Tee Craft came out from Michaels Boat Service whom had a small crane and moved Kathy's boat away from her house.

Johnnie and Monaca came by advised FEMA may provide financial assistance and also advised Johnnie had helped Monty my tenant in the Redlands move tree limbs from his gate and also provided power to Monty's home from his RV.

The Coast Guard contacted me by text message and advised that my EPIRB had gone off early Sunday morning and they gave me the coordinates 24 degrees, 25.1 minutes north and 081 degrees 25.5 minutes west. I checked this put it about fifteen miles SSE of Sugar Loaf Key in deep water.

Ramon and I took all furniture out to large trash pile in the road in front of the house. We continued to get Dick to charge or mobile phones.

Friday morning, we still have no power or internet. Fresh percolated hot coffee this time for us thanks to Kathy and cereal thanks to Igor with milk. Took Ramon to get his phone plan updated at T Mobile and we ate at the Juice House a few doors down. Got into an argument with Ramon about his very expensive plan he wanted twice the cost of my and my wife's plan.

We then went to borrow a large metal jig saw from his friend so we could cut some rebar in the cement shoulder previously holding my floor pavers now all ripped up. Dick kept charging our phones thankfully.

That day we worked hard shoveling off all the gravel about a foot thick over all my tile floor area downstairs. I made a bridge with concrete floor tiles over the concrete shoulder and we wheel barrowed the gravel over and into the large holes near the dock. Ramon acted disturbed all day doing this work it was hot and I think he felt he had done enough for just room and board. We worked late and all the stores closed at 5:00 p.m., so Ramon bought expensive beer from Tom Thumb.

That night Ramon called his mother and father in Tahiti. After talking to his mother, he walked off to talk to his father and we never saw him again. Next morning the beer was gone from the ice chest and also gone was Ramon and his cat Nate.

Saturday morning Carlos came over and we looked for the *Borneo Princess* on a live satellite camera obtained by logging on to NOAA hurricane Imagery.

We looked zero to four miles out from the Seven Mile bridge and all along the bridge and found sunk boats even a catamaran sailing vessel but not mine. We then drove to my Redlands home to get my seventeen hundred a month rent from Monty and to discuss what to do with Johnnie Rodriguez. Monty felt uneasy with Johnnie living on his property and his daughter Jasmine staying alone during the day there even felt more afraid. We decided to tell Johnnie to leave and take his RV with him.

Story of Borneo Princess and Hurricane Irma

Came back and went to Winn Dixie and got ice, hamburger, Sprite some canned food and water melon.

Batit sick of the mess and also not aware Ramon was no longer with us had made plans to stay with a friend in Fort Lauderdale for a few days. She also attacked me verbally, blaming me for the sinking of the *Borneo Princess* at which point I blasted her, telling her to leave the house right then and there.

Dick and Kathy asked if I wanted to share two laborers the next day Sunday they had planned to drive and pick up at Homestead. I agreed.

That night I went to Carlos's home and had a few drinks and made hamburgers. I rang Kathy who informed me that power was now on so I left for home. Batit had responded to my demand and had already left by an Uber Taxi. She did however turn the AC on and closed all the windows and doors before she left.

Oh my God how good it was to feel air conditioning again, have a refrigerator and cook on a stove.

Sunday, I had two Pop-Tarts, some of the food Igor left, and a hot fresh coffee. I turned all my downstairs power on except the 220-volt stove, which Michael and I could not find a circuit breaker for. Dick returned one of the two gasoline cans he borrowed.

I bought some gloves for the new workers but was never to see them do any work for me they spent the day with Kathy and Dick. I bonded a little with Masey one of the two black Haitian workers and said I would ring him to come help me when I could see my way through the mess.

I completed shoveling off all the one-foot-thick sand and grit from my tiled areas outside. Filled six wheelbarrows and took it over my bridge to dump near the dock. Then swept and washed down tiled areas.

Carlos came over and we drove down sixty miles and across the Seven Mile bridge noticing the complete destruction of homes and especially trailers there and also looked out over the water for the *Borneo Princess*. With binoculars we saw something sticking up out of the

115

water about a mile or two out which we thought could have been my mast. Later investigations showed it to be a navigational tower.

Later Carlos took me to lunch and we shared a Dolphin, now called Maui Maui, sandwich. That night at 5:00 p.m., he was invited to a free steak Bar Be Q but BYOB (Bring your own beer) at the location of the previous Snappers Restaurant. I provided the beer and off we went. Met some of his friends; one was Connie, queen of the Key Largo renter's association, and we all swapped hurricane stories and drank a lot.

Batit rang and woke me at 1:00 a.m. Monday morning unhappy and had lost USD 6,000 at the Hard Rock Casino. I was so happy she called I did not get mad at her losing her money.

Monday, I woke up a little sad I knew I had no one here to help or keep me company. I cooked some eggs for a good breakfast. I went down started picking up some of the 250 heavy 20-pound concrete tiles that were strewn all over the backyard. I could only lift and stack about twenty-five of these a day in the humid heat before I got exhausted. I decided I needed help so rang Masey and he agreed to drive down on Wednesday morning and work for $180 a day including travel.

I was going to have to cut some rebar to remove the concrete boarder that was laying across the area so I went to Harbor Freight in Homestead and unbelievably bought a hand grinder on sale for $20. I also picked up there some grinding discs, a 100-foot heavy duty extension cord and a heavy-duty push broom. I had to go to Home Depot to get a strong steel wheel barrow. I spent about $180 but saved $70 compared to AKI prices in Key Largo. I also went to the Family store in Florida City and bought two indoor couches, two outside couches, two bedside tables, two chest of drawers and a wicker chair.

I watered heavily all the trees and plants for twenty minutes each that had been flooded with salt water. Tuesday, I finished watering all the plants we had not seen rain since the hurricane nine days ago.

I found a guy on the phone and agreed to pay him $100 to transport all my furniture to Key Largo. The driver also helped me carry everything inside and also take some furniture outside. I wanted so much to make the house look good for Batit when she returned. She agreed for me to come pick her up Wednesday night.

Airbnb contacted me by text and I agreed to let them cancel the booking for 8th October.

I contacted Steve at Wire Nuts Electric to get help with my shore power connections but he refused saying he had other priorities which was to get power to people's homes.

Wednesday was a great day Masey came and shoveled off all the concrete tiles and stacked about 200 of them making the backyard look a lot cleaner.

I picked up Batit that night in Fort Lauderdale and we had joyful reunion in our beautiful and now air-conditioned home.

23.1 The Funeral

Next day, Thursday, we get up late I cut up the dead Frangi Pangi tree in the backyard and take it to the trash and remove all the limbs from the Royal Poinciana tree and stand it up holding it up with a rope.

We then go to my mother's house to stay the night and attend Victor Kauffman my step father's funeral the next morning. Victor had passed away in the hospital a few weeks before the hurricane hit. I had debated for weeks not going. I did not like Victor, and over the last five years, I felt he had been unfairly rude and cruel to me and Batit for no apparent reason except he was old and sick. Things culminated when he invited us to dinner and we had to drive four hours in heavy traffic and arrived a little late, at which point he blasted us over it. We went to the Outback, a supposed Australian restaurant, and when I started talking Aussie, Victor started to get mad, at which point I lost my temper at him for the years of abuse he had shown us and he would never talk to me again as he would die within one year.

I also had not been on speaking terms with my sister Lynn for many years stemming from silly reasons resulting in her rude husband, Jack, sending an email that they never wanted to talk to us again. Then they and my mother failed to attend my wedding in 2013 and every year took my mother and Victor to their house for every Christmas and Thanksgiving without ever once calling us.

My brother Bill I also had a recent falling-out when the company where he had worked for sixteen years ran out of money, and since I owned most of the shares, he expected me to pay him a couple month's salary he was owed, and I would not. In fact, he owned 6% of the company, which was in debt and his share owed was twice what was owed him.

We were met with open arms by my mother; evidently, without Victor, she was able to think and feel on her own. She bought hamburgers at Burger King earlier that day, which we ate. I called Lynn and Bill just to say a simple hello and would see them tomorrow. That night I slept in Victor's bed alone Batit refused to sleep there… something about his spirit still around, so she slept in the living room on the couch. That night my stomach burned not knowing if it was the hamburgers, the pistachio nuts I consumed on the four-hour ride up, the rye whiskey I had drank that night, or was Victor's ghost screwing with me. I could not sleep well and had constant bad dreams, including one that the house was sinking and I was trying to put the plug in as if it were a large boat.

The next day we got up, my mother made her standard breakfast cereal, coffee, and a muffin. I put the address in my phone as I had MAPS and easily found a route and was sure it was correct as the name of the church was also displayed at our destination. MAPS had me deviate a little, probably to avoid heavy traffic, and my mother went crazy not trusting a computer and proceeded to start changing the address as well. I had to yell at her to sit down and be quiet.

We made it to the Our Savior Lutheran church in quick time much to my mother's amazement, as evidently, she and my sister could not find it easily the day before. We met outside a few people: Ray Clark

Victor's nephew and his wife and his son Leroy Clark and wife. Leroy and I spoke at length about the Keys, where he also lived. I remember they had repaired Victor's roof a few years back, and I asked them if they could look at my roof as I think I lost a few tiles during the storm.

There was a very nice ceremony with the local pastor saying some words, a few hymns, and then my sister, brother, and I stood up and said a few words about our lives with Victor. Everyone had fond memories and he had been a great guy to me for forty-five years out of the fifty I knew him. They had many framed photos of family and friends. We noticed the one my mother had kept of my family reunion on Christmas 2009 in Australia—of Batit and me, Bill, Sari and Willie, Sean and Yusi, Melissa and Maverick and Victoria and Dylan—was not there nor was it at her home. It was found two weeks later by my mother with no frame. Someone sick obviously planned on destroying it Vic, my mother or sister.

That night I took everyone out to an all-you-can-eat Chinese buffet. Bill had taken Lynn and his family in his car. Later Lynn kept calling as she had lost her keys, which I found and she came to get them. Later Batit and I slept in my mother's other bedroom in the small double bed with canopy. We slept soundly.

Next day Bill and Sari came over and we said goodbye. Bill stayed and fixed her mailbox. We drove to Jensen Beach the next morning to visit with Van, a forty-year-old pretty Vietnamese female canvas artist from California Batit had met at Lonnie's who had now hooked up with a twenty-year-older man named Tom living in Jensen Beach. Tom was very intelligent, outgoing, and fun besides being a successful businessman and we immediately bonded. We drove home later glad to be in our comfortable home.

23.2 The Clean Up Continues

The next week, we still had no internet and so I was making frequent trips to The Juice House to check and reply to emails. I would go to Comcast every week just to remind them of my situation and they promised me Wednesday it would be working. I would do odds and ends with the endless cleanup. On Tuesday I had arranged for Macey to come again. I also had heard from Joe Hardy, who had helped me on my boat for free in return for free storage of his boats at Redlands and for his personal effects in my box truck. It was supposed to be for a few months but had now been five months. Joe came with a bobcat and two helpers and I agreed to $120 per hour.

They all worked hard for three hours and returned much of the gravel that had been washed away from the concrete dock. In addition, two fences were removed and parts of my broken concrete dock demolished. They later helped me take my very large broken aquarium out front to the trash pile that I had kept outside to hold lobster and fish alive. We achieved the most cleanup that day.

Joe and his two friends and the older man's wife were staying in a $200 per night hotel, and so I offered them my bottom apartment for $70 per night but had no stove and needed a queen mattress and box spring.

Wednesday, I went and got the mattress and box spring at the family store in Florida City. I had only paid $80 and put them on the roof of my Ford Explorer SUV and drove to Key Largo, hoping for no rain.

When I arrived in Key Largo, Joe had a truck and his sixty-year-old-something friend's car parked in my front yard. I asked his friend to please move his car to the vacant lot across the street at which point his friend got angry. I went inside and the friend's wife, Kareen, was trying to clean up but water was draining on the floor and her stepson, Dan, did a quick job of fixing it. Evidently, the drain had moved during the storm surge. We also brought in the mattress and straightened the place up and brought in the sofa bed that was outside.

The subject of their large Labrador Retriever dog came up. I had agreed with Joe it had to stay outside. She did not like that idea, and I said okay, it can come inside but they must not let it on the furniture and I don't want it peeing on the floor. At that point, she left and I never saw her or her redneck husband again. I offered the apartment to Joe and

Dan, the rednecks' son that had fixed the sink drain and helped me move furniture for $50 per night. Batit cleaned the apartment for four hours.

They came back and stated for four nights, never to have be seen again. They paid the $200, but I was angry they had just left and so told Joe by text to get his boats out of Redlands and empty my truck of his belongings. Saturday Batit and I worked hard for many hours opening all the storm shutters and washing them then closing and re-securing. Sunday Batit and I drove to Key West and had dinner at Hogs Breath Saloon. The food was terrible, the beer expensive. Upon our return, no Joe and never saw or heard from him again. On Tuesday morning I met Johnnie at The Juice House and he agreed to help me get and empty the truck. We put a lot of things in the trash pile and kept fishing rods and large boxes of Nutriment foods. The Nutriment foods were a complete set of foods for three weeks including cereals, pizzas, meat stews, cookies and candy bars wrapped in sealed plastic. I also agreed to pay Johnnie $100 to take and fix my gasoline powered water blaster unit.

Nothing much seemed to happen on the clean up early in the week except me shoveling and raking a little gravel. My Tee Craft boat rub rail had been beaten up on the piles so I took a few hours on Thursday to install the spring poles that keep it off the wooden piles. Comcast did however come out that we Wednesday as promised and installed my cable underground in one of the two spare conduits I had installed with my underground electric years before but still not working there was a cable break at the pole.

I talked with Alfred an electrician from Wire Nut Electric working at Igor's on a boat lift of all things and told him about no 220-volt power down stairs. He needed authorization from his company so knowing all the phones were tied up and previous problem I had had with the company I decided to go meet the company face-to-face.

It worked. Marnie, the owner's wife, authorized Alfred to fix my power. He found the breaker upstairs, but it was faulty sticking sometimes. He then replaced the receptacle as it was shot.

We tested the stove; it was no good, so we took it out to the trash pile. During this time Comcast came out and repaired the broken cable at the pole and we had TV and internet. No more going to Juice House or sitting in the house thinking. I watched some of my favorite movies Friday and Saturday night *The Deer Hunter* and *Jeremiah Johnson*.

The county trash collector came out late Friday in the afternoon and with rubber tracked vehicles fitted with grab devices started picking up the trash and placing in numerous trucks that showed up. They would not touch the refrigerator and my stove as that is picked up separately. My 200-pound large aquarium also left behind. I had bonded a little more with Dick that day as I swept up and he shoveled debris in the street to the pick-up vehicle.

On Saturday morning Leroy and Ray came over and inspected my roof to report only seven broken tiles and one ridge capping missing and estimated a few hundred dollars to fix. The FEMA people also came out on Saturday and another on Sunday to register my damage claims. I did not expect to get any help as they were not covering docs or any rental properties which downstairs could be considered to be. There were low interest business loans available I would learn about. Saturday my aquarium was removed finally attributed much to my constant polite nagging them every day. They were all worried it would break up and I assured them it would not, and if it did, I would pick up all glass. Johnnie came around and returned my high-pressure water blaster.

The apartment would need a new stove, refrigerator, bathroom vanity, kitchen sink cabinet, the back aluminum weather shutter pushed out and the stucco outside repaired. The next guests were scheduled for December so had plenty of time. My dock needed a lot of fill that had been washed away, repair of the electric lines to the davits, and a dock power receptacle, repair of a water line, repair of my two boat lift davits, and repair of my 40-foot-long lower dock, replacement of a light pole and two chain link fences. The real money now had to be spent but I would figure out a way to keep costs down.

On Tuesday I get a quote of USD 28,000 to fix my dock which I thought was outrageous. A few months later I met by accident a Mr. Eric Wall a thirty-five-year-old construction carpenter at the local KLI hardware store as I overheard him talking about concrete work. I figured I needed a 2-feet high concrete wall to hold back the soil and also connect the floor joists to wood beams connecting my large wood boat mooring piles. I also wanted a white PVC deck and the best treated wood and 316 stainless steel bolts, hangers and screws you could buy. Eric helped me finalize my design and also did most of the work resulting in a total materials cost of $3,800 and for his labor about 100 hours x $30 per hour or $3,000 for an all up price of around $6,800 a far cry cheaper than $28,000 or about 22% of. I helped Eric but I figured my involvement as just exercise.

The new deck and repairs to the property ended up costing about USD 16,000 not including my time. I was making weekly driving trips to Islamorada an about 15-mile drive South to FEMA (Flood Emergency Management Association) headquarters because I heard they were handing out financial assistance. I filled out applications for assistance and found out two other persons fraudulently had done the same for my property. After one trip down I produced the property tax file receipts in my name which must have worked because to my surprise one to two months later after I had given up all hope of FEMA assistance, while making a trip to my bank discovered they had made a USD 18,000 payment.

24.0

The Aftermath

24.0 The Aftermath

Toni Marchetti on October 10 tells me Sam has kept his boat now a month after the hurricane and is demanding unreasonable fees to release his boat and asks me to intervene, so I advise Sam by text. In middle October, Sam rings to say he has made USD 500,000 pulling and storing boats. In the 18 October local paper, I read how the District Attorney is threatening to put Sam in prison if he does not refund much of the money he had inappropriately taken from boat owners he blackmailed and held their boats for ransom. Steve Lavine advises Sam refund Toni $2,500 of the $5,000 he had taken.

A few weeks later, I read in the newspapers Sam's boatyard office and two boats had burnt to the ground. I texted Sam, although we don't talk anymore, and he said one of the boats' dehumidifiers next to his office caught fire.

A guy rings me from near Fort Pierce says he has found my EPIRB floating, but I lost his number and did not hear from him until ten months later when NOAA finally advised me someone had found my EPIRB and wanted to know if anybody was hurt and I advised no and then requested the finders number. His name was Mike Gasier and he lived in Stewart, Florida. We later arranged to meet almost a year after the hurricane and he would return the EPRB and I offered to print his story of finding it, which is the Prologue of this book.

The company Reef rings me two weeks later to say they spotted my life raft canister up against the mangroves just north of South Cut into Largo Sound.

We would continue to fix the bottom apartment and rented it out three times in November 2017.

About a year later, I would get my power pedestal re-installed by Wire Nuts Electric.

Almost a year to the date from the storm I was sitting in a small café called BJ's in Key Largo and am sipping an ice tea when I pick up off the counter in front of me a copy of Keys Weekly and read "Eight Things to Know about Irma" and almost verbatim it states the following:

THE STORM Irma made landfall on Cudjoe Key at 9:00 a.m., September 10 at a category four intensity with sustaining winds exceeding 130 MPH. It moved North along the west Coast of Florida bouncing off and hitting the coast seven times. Meanwhile 100 miles to the north in Key Largo that night about thirteen hours later we would get a 10-foot storm surge and winds of 115 mph with 6-foot to 12-foot waves depending upon where you were breaking on to land

THE HOMES Nearly 40,000 homes were affected and approximately 1200 were completely destroyed with most occurring in Marathon and Big Pine Key. Most of those completely destroyed were mobile homes or older homes built prior to more stringent building standards now in place. Johnny Rodriquez would make a lot of money over USD 100,000 getting for free, transporting and selling mobile homes in Key Largo and Miami.

THE DEATHS According to Monroe County medical examiner Dr. Michael Steckbauer, there were seventeen hurricane related deaths with three caused directly by the storm and fourteen by undetermined means starvation, drowning, and heart attacks. Luckily there were no more.

THE DEBRIS According to county spokesman Cammy Clark 2.5 million cubic yards of debris were removed including vegetation and demolition waste. About 20,000 appliances were removed from Keys roads where many had been for many months. Much of the vegetation and appliances were piled as high as 50 feet and stored Bay Side at Key Largo. Some of the vegetation was burned most of the white goods went to the mainland for recycling including my refrigerator washer, dryer and electric stove.

THE BOATS Clark reported as many as 1600 damaged or destroyed boats were removed from the Keys. Two thirds of the 300 boats docked at the Boot Key Harbor City Marina in Marathon were destroyed along with many others that either drifted away or sank. The Florida Wildlife Conservation Commission, the US Coast Guard, The State Department of Environment Protection and the US Environmental Protection agency worked together to salvage and remove boats and especially their fuel. All of the boats in Key Largo on the open water on boat lifts or docked were destroyed along with the boat lifts and then many simply washed crashed into the adjacent homes.

FEMA FOR THE HOME LOSS AND DESTRUCTION As of September 4, 2018 the Federal Emergency Management Agency (FEMA) has paid out $334 million to home owners and businesses and in addition provided $180 million in low interest loans to home and business owners to rebuild. That represents an average grant or loan of $12,850 to each of the 40,000 home owners. There are still 104 households living in temporary FEMA trailers.

FEMA FOR THE COUNTY As of August 2018 Monroe County has lodged over $39 million in county related damage claims to FIMA for reimbursement. Only $8 million was made available to the state for distribution to the county and as of Aug. 20 only $2.3 million has been received.

CANNALS About 1100 cubic yards of marine debris has been pulled from 103 Keys Canals by a Consortium of subcontractors including ASAP Marine Contractors of Tavernier, Arnolds Towing of Stock island, and Tetra Tech Diving managed initially by DRC Environmental Services and more recently after DRC pulled out by Adventure Environmental. Monroe County has received a grant of $49.2 million from the US Department of Agriculture's Natural Resources Conservation Service (NRCS) to pay for the work. There will be six crews working on fifteen barges that just started on August 17, 2018 and will continue for another seven months.

OTHER Hurricane Irma killed a total of 129 people on its rampage through the Caribbean and Florida. The storm was the most searched topic globally in 2017.

Carlos, his wife Luda, Ramon, and myself drove to Marathon on September 9, 2018 to commemorate the one-year anniversary of the fateful day of our rescue. Johnnie Rodriguez in prison and Mike Maffri not wanting to talk to us were the other two survivors.

We had a pleasant picnic brunch on the bayside feeding pasta to many nice snapper fish and a late pizza lunch in Marathon toasting each other with glasses of fine wine.

25.0

Photographs

Figure 1: Installing Stiffeners in the Rolled Pontoon Plates 2011

Figure 2: The Structure and The Second Bow Being Added

Figure 3: Fitting the Aluminum Windows

Figure 4: Sean inside the Starboard Pontoon above a SS water tank

Figure 5: The Mast Support Structure 20 Tonne
Load Any Direction 24 ft. Above

Figure 6: The Protected Propellers, Kort Nozzles, Water Access Stairs

Figure 7: Launching Changing Shipyards 2013

Figure 8: Successfully Launched AT Meranti's Ship Yard

Figure 9: Painting and Handrails Going On

Figure 10: Access to Saloon Roof with Inorganic Zinc Primer No Top Coat Yet

Figure 11: View from Port Stern Handrails towards Entrance

Figure 12: Stern View First Dry Docking
to Repair launching Damage

Figure 13: Second Dry Docking at Maranti's Ship Yard to Repair Damage from Barge Collision

Figure 14: Port Side Master Suite with Bathroom on Right

Figure 15: Saloon Flooring and Trim

Figure 16: Port Engine Room

Figure 17: Wheel House

Figure 18: Sean with Dion and Dody

*Figure 19: Sea Trials Captains Simon Thompson
and Joe Dyll (background)*

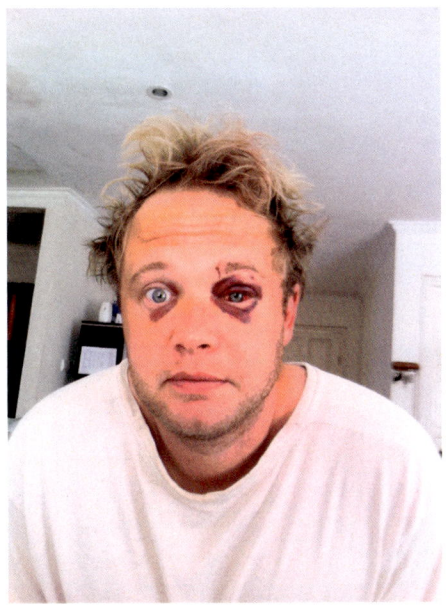

Figure 20: Sean Next Morning after Bad Night out Missed the Trip

Figure 21: Completed Vessel Enroute to Singapore with Ryan, Dodi and John on Deck, Dion on Dive Platform Waiting to Receive Clearance Papers from Sean and Captain Simon in Bridge

Figure 22: Enroute to Singapore Photo taken By Sean

Figure 23: Our Security Ryan and Dion with Machetes

Figure 24: Galley and Pantry to Right

Figure 25: Another View of Saloon Looking Forward

Figure 26: Three Days Out from Balikpapan
Kort Nozzles Recovered on Deck

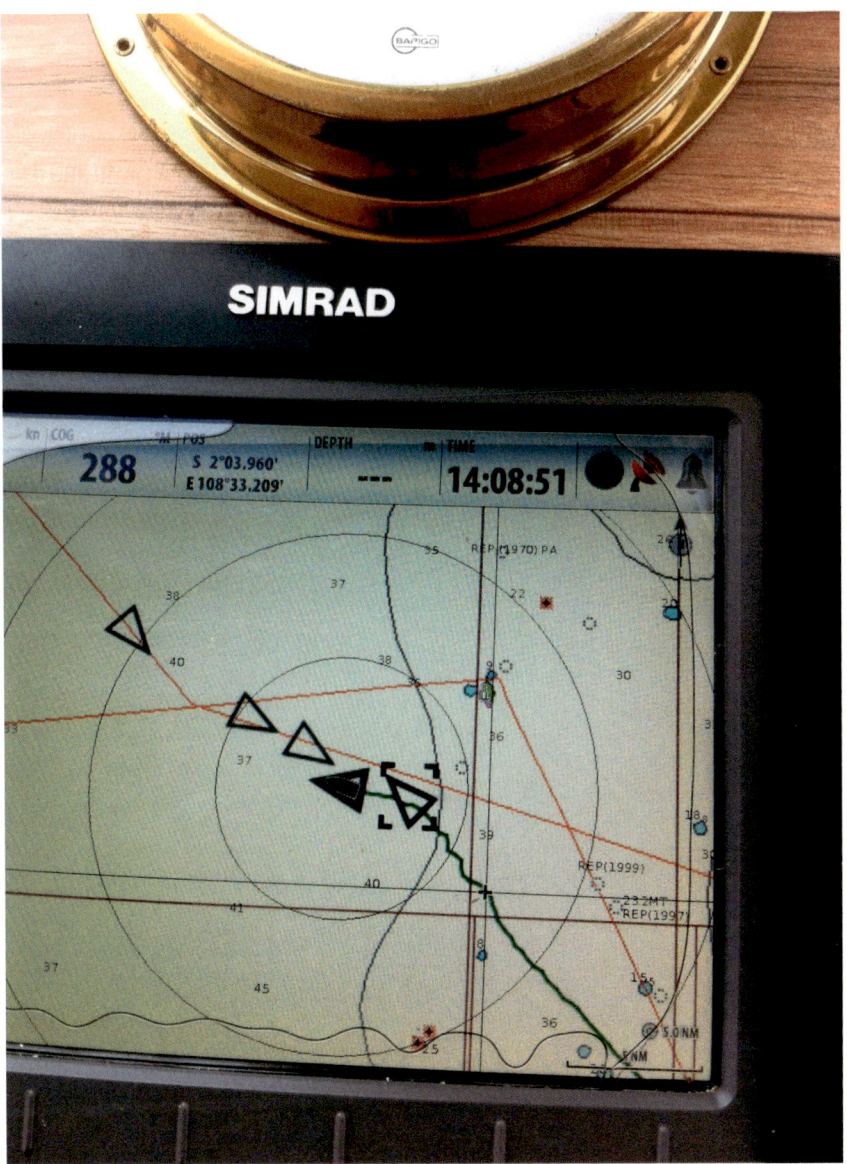

Figure 27: On the AIS Display We are Black the Other Four Triangles Represent Large Ships Three Coming at Us from Ahead and One from behind All on A Collision Course

Figure 28: One OF the Ships a Six Hundred Foot Freighter Could Run Us Over and Never Know It

Figure 29: Batam the Princess Finally Would
See Other White Luxury Boats

Figure 30: On the Ship Bound for Fort Lauderdale Notice the Difference Between Standard Boats Out There

Figure 31: Finally, Behind the House February 2017

Figure 32: At the Sea Aquarium for Fox Contract

Figure 33: The Film Crew Filming Chum

Figure 34: The Princess Covered in Fox Studios Wax Scum
Which Could Not be removed

Figure 35: Returning Home from Fox Dirty and Bent Up

Figure 36: Port Side Shaft, Coupling and Prop in for Repair

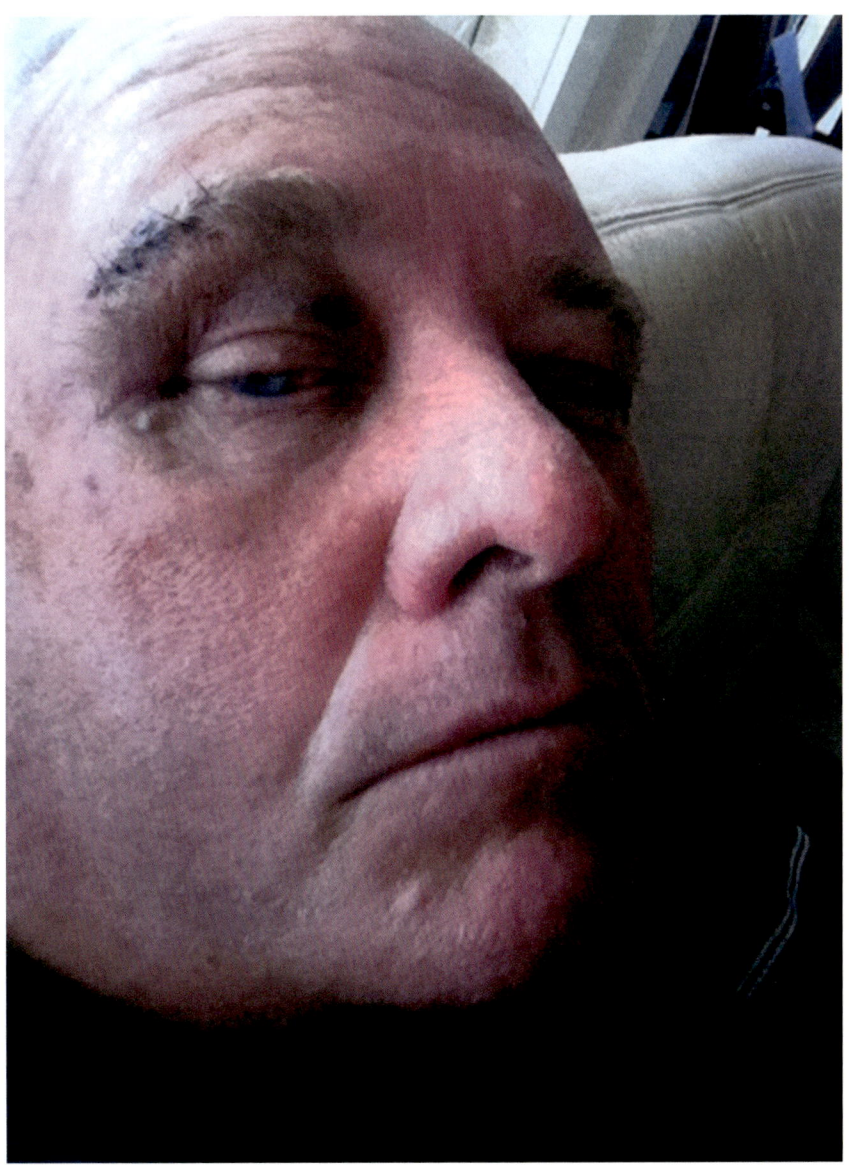

Figure 37: Me After Getting Sewn Up from
a Fall Coughing with Pneumonia

Figure 38: At Rodriguez Key and Installed Back Deck Tarp

Figure 39: At Home Again Cleaned Up with New Protective sail Cover Ready for Next Charter to Cay Sal

Figure 40: Carlos Hitch Hiking from Key West to Key Largo

Figure 41: Our Last Stove Cooked Supper Nurse Shark Stew Caught by Johnnie and Prepared by Ramon on the Night of the Storm

*Figure 42: The Water Is Rising, Storm Surge 10 feet,
Wind Blowing 140 knots, 4 ft Waves*

Figure 43: Mail Boxes Gone Next Our Couch Downstairs Disappears Out Gate across Street Then to Canal in Back

Figure 44: A Sign That We Used to Have Out to Slow Boaters Ends Up in the Front Yard Up Against the Fence "Slow Down Dad End"

Figure 45: Out Go's he Table and chairs Batit
so Carefully Tied Together

Figure 46: With No Electricity Luckily Ramon Finds
Our Barbeque, Drift Wood and Cooks Meals

Figure 47: My Dock

Figure 48: My Neighbors Boat Wrecked into Her House

Figure 49: Another Neighbor's Boat into His Swimming Pool

John Lincoln

Figure 50: Another Neighbor's Wrecked Forty Foot Tierra

Figure 51: Another Neighbor's Wrecked Wood Dock

Figure 52: My Backyard Pavers Everywhere